CHOICE-BASED CONJOINT ANALYSIS

MODELS AND DESIGNS

DAMARAJU RAGHAVARAO
JAMES B. WILEY
PALLAVI CHITTURI

CRC Press
Taylor & Francis Group
Boca Raton London New York

CRC Press is an imprint of the
Taylor & Francis Group an **informa** business
A CHAPMAN & HALL BOOK

Chapman & Hall/CRC
Taylor & Francis Group
6000 Broken Sound Parkway NW, Suite 300
Boca Raton, FL 33487-2742

Printed in the United States of America on acid-free paper
10 9 8 7 6 5 4 3 2 1

International Standard Book Number: 978-1-4200-9996-6 (Hardback)

Library of Congress Cataloging-in-Publication Data

Raghavarao, Damaraju.
Choice-based conjoint analysis : models and designs / / Damaraju Raghavarao, James B. Wiley, Pallavi Chitturi.
p. cm.
Includes bibliographical references and index.
ISBN 978-1-4200-9996-6 (alk. paper)
1. Conjoint analysis (Marketing) 2. Consumers' preferences. I. Wiley, James B. II. Chitturi, Pallavi. III. Title.

HF5415.2.R24 2011
658.8'343--dc22 2010007126

Visit the Taylor & Francis Web site at
http://www.taylorandfrancis.com

and the CRC Press Web site at
http://www.crcpress.com

Contents

Preface

Individuals reveal their preferences for decision alternatives through choice. Understanding how individuals develop preferences for decision alternatives is a fundamental task that facilitates effective management in government, for-profit, and not-for-profit business. Conjoint analysis (CA) and discrete choice experimentation (DCE) are tools developed since 1960 for understanding how individuals develop preferences for alternatives.

A major subset of issues related to the formation of preferences occurs when alternatives may be described by attributes, and the objective is to infer the values respondents attach to attribute levels. This book reviews experimental designs useful for inferring the values that individuals attach to attribute levels. It will be useful to researchers in psychology, economics, geography, marketing, tourism, human resources administration, health administration, and the political and social sciences. It will also be useful to statisticians and combinatorial mathematicians for finding different applications for the wonderful configurations they developed with other applications in mind.

While the emphasis here is on experimental design issues, we also discuss the concepts and inference methodology in the text to provide an idea of the underlying philosophy of the conclusions drawn from data. To understand the estimation and inference material, a minimal mathematical background is required (i.e., familiarity with concepts such as expected value, variance and dispersion matrices, normal and chi-square distributions, unbiased estimators and standard errors, testing statistical hypotheses, and matrix manipulations).

We note that there are important topics in CA/DCE other than experimental design that are either not covered or not emphasized in this book. For an additional basic introduction to CA, we recommend the work of Louviere (1988) and Orme and King (2006). For an introduction and overview of DCE, see the material presented by Louviere, Hencher, and Swait (2000). Although there are several ways that data from CA/DCE experiments may be analyzed, in this book we assume that the results of CA/DCE experiments will be analyzed using a multinomial logit model. For additional coverage of estimation procedures, we recommend Train's (2003) work. We also do not emphasize statistical properties of designs, such as A- or D-optimality. For coverage of designs that emphasize these issues, we recommend the study of D. Street and Burgess (2007).

Chapters 1–3 provide background material for the subsequent chapters. Chapter 1 introduces the notion of revealed versus stated choice experiments and describes CA and DCE experimentation. Some historical background and examples are provided. The multinomial logit model and estimation are reviewed. Chapter 2 reviews experimental design material that will be used in subsequent chapters. Throughout the book, the concepts relating to experimental design are explained by suitably selected examples so that researchers with minimal mathematical and statistical background can understand them. Chapter 3 provides examples of "generic" experimental designs that are commonly used in CA and DCE experimentation.

The emphasis in this book is on developing practical designs for specific classes of problems. For example, the attributes in CA/DCE experiments often (if not usually) are economic ones (i.e., benefits or costs for which the value associated with attribute levels monotonically increases or decreases with level). With such attributes, it is possible for a multiattribute choice alternative to be better or worse on every attribute than another choice alternative. Chapter 4 deals with the design of experiments that do not have such dominating or dominated alternatives. It also illustrates the use of orthogonal polynomials to describe the relationship between attribute levels and preference.

Chapters 5 and 7 discuss designs relevant to situations for which the CA/DCE experiment will be administered by a computer (e.g., when it will be administered over the Internet). Typically with computer-administered studies, choice alternatives are presented on computer screens. This limits the amount of information that may be presented on one screen. Chapter 5 discusses approaches subsetting the attributes or levels presented. The material in Chapter 5 is also relevant to Chapter 4 material. The procedures of Chapter 4 may result in large choice sets, and the subsetting procedures of Chapter 5 may be used to reduce the choice set sizes presented to respondents.

Computer administration allows for the possibility of having the presentation of subsequent choice alternatives depend on prior responses. Chapter 7 discusses strategies for sequential experimentation. Sequential experimentation allows the researcher to sequentially diagnose the model used by individuals to combine attribute information into expressions of preference and choice. The material in Chapters 5 and 7 is also relevant to exploring behavioral issues relating to the choice of experimental design (e.g., exploring the consequences of choice of profile design or order of presentation on observed choice).

The attractiveness of a choice alternative may depend on the other available choice alternatives. Chapter 6 discusses designs relevant to "availability" problems. Availability designs are appropriate when the attractiveness of an alternative depends on the availability (absence or presence) of other alternatives in the choice set. When alternatives are characterized in terms of attributes, the level of an attribute of another alternative may affect preference for an alternative. For example, the effect of a price change will depend on whether competitors match the price or change other attributes in response. Chapter 6 discusses alternative and attribute cross-effects designs and the relationship between availability effects and an alternative way of capturing the effect of one alternative on another: interaction effects in analysis of variance (ANOVA) models. Availability designs are also relevant for studying portfolio choice. *Portfolio choice* refers to situations where respondents choose more than one alternative in a choice set or choose from choice sets for which alternatives are themselves bundles of alternatives. The chapter concludes with an illustration of the design and analysis of a portfolio choice problem.

Finally, Chapter 8 discusses approaches to experimental design in which constraints are imposed on the levels of attributes. For example, there may be costs associated with attribute levels, and a budget constraint may limit the alternatives that can actually be offered. Mixture designs refer to when a fixed resource, such as total cost, must be allocated to a set of attributes and the problem is to determine the optimal allocation of the resource among the attributes. For example, a developer of a new product or service may want to develop a product that can be sold profitably at a given "price point." The design problem is to determine the most preferred combination of attribute levels, subject to a budget constraint for the project as a whole. A more general problem is to determine not only the optimum allocation of a resource but also the optimum level of the resource, that is, the most profitable price point and the optimal allocation at that level. Mixture/amount designs are relevant to this problem. Chapter 8 discusses and illustrates mixture and mixture/amount designs.

Aside from the references provided in the text, there are several computer programs that may be used to generate experimental plans. The Sawtooth Software Corporation provides a variety of products for implementing CA and DCE studies (Sawtooth Software, 2009). The ORTHOPLAN module of SPSS (2008) provides the capability of generating main-effects models and analyzing the resulting data. SAS FACTEX and OPTEX modules (2008) provide extensive capability for constructing and analyzing designs, as does the Design-Expert program provided by Stat-Ease Incorporated (2008). The main emphasis of this book, however, is to provide designs mainly for DCE, and analysis of these designs is of secondary interest as different researchers analyze data collected using the designs differently. We provide detailed coding for design matrices for standard designs while sketching analysis for complicated ones.

D. Raghavarao

J. B. Wiley

P. Chitturi

About the Authors

Damaraju Raghavarao received his Ph.D. from Bombay University, India, and is currently the Laura H. Carnell professor of statistics at Temple University (Philadelphia, Pennsylvania). He was awarded the Khan Bahadur H. M. Malak Gold Medal in 1957 for his performance in the Master's Examination by Nagpur University, India, and received the Paul W. Eberman Research Award in 1995 from Temple University for his research accomplishments. One of his major contributions is the monograph *Constructions and Combinatorial Problems in Design of Experiments*, published by Wiley in 1971 and Dover in 1988. He is a Fellow of the Institute of Mathematical Statistics (IMS), a Fellow of the American Statistical Association (ASA), and an elected member of the International Statistical Institute (ISI).

James B. Wiley has a Ph.D. from the University of Washington and is senior Cochran research fellow in the departments of marketing and statistics at the Fox School of Business, Temple University (Philadelphia, Pennsylvania), and a visiting scholar at the University of Western Sydney, Australia. He is the outgoing business and market research track editor of the *Journal of Business Research (JBR)* and is on the editorial review board of *JBR* and the *Journal of Food Products Marketing*. Previously, he was the Australia/New Zealand regional editor of the *Academy of Marketing Science Review* and a book review editor of the *Journal of Marketing Research (JMR)*. Past positions include professor at Temple University; University of Alberta, Canada; and Victoria University, Wellington, New Zealand. He was a Fulbright scholar in the Republic of Slovakia and held visiting positions at the University of Florida, Gainesville; the University of New South Wales, Sydney, Australia; and Hong Kong University of Science and Technology. His publications have appeared in the *Journal of Marketing Research, Marketing Science, Psychological Bulletin, Journal of Multivariate Behavioral Research, Utilitas Mathematica, Journal of Statistical Planning and Inference, Communications in Statistics*, and other journals.

Pallavi Chitturi received her Ph.D. in statistics from the University of Texas at Austin. She is an associate professor of statistics at Temple University (Philadelphia, Pennsylvania). Dr. Chitturi's research interests are in the areas of design of experiments, quality control, and conjoint analysis. Dr. Chitturi has made several research presentations at conferences, such as the Joint Statistical Meetings and the Conference of Texas Statisticians. She has published in statistics, marketing, and quality management journals. Dr. Chitturi is a recipient of the Andrisani-Frank Teaching Award. Dr. Chitturi has received grants from the U.S. Navy's Naval Logistics Readiness Research Center and one grant from the Office of the Vice-Provost for Research at Temple University.

1

Introduction

In real life, people reveal their preferences through choices. The aggregate of choices constitutes the demand for goods and services, the vote for political candidates, and many other phenomena of interest. Understanding how changes in the characteristics of alternatives affect preferences for them is important in many fields in which predicting human choice is of interest. Such fields include psychology, economics, environmental science, geography, management, marketing, political science, recreation, and transportation.

This book deals with situations for which choice alternatives may be described in terms of their components or attributes. For example, price is an attribute that influences choice of an automobile Interest may be in several price points, or levels, such as $21,000, $26,000, $31,000, $36,000, and $41,000. Other attributes might include brand name (with levels: Ford Taurus, Chevrolet Malibu, Mitsubishi Lancer, Volvo C30, Honda Accord); number of doors (two, four); size of engine (four, six, or eight cylinders); and type of transmission (manual five speed, automatic five speed, manual six speed, automatic six speed). Under certain conditions, it is possible to infer the partworth (or part utility) of the respective attribute levels by regressing information about product attributes on sales or market share. Such data are referred to as *revealed preference* (RP) data.

The problem with making inferences about partworths from RP data is that all the data required for estimation of partworths may not be available. For example, some combinations of attribute levels may not be observed. It may not be economically or managerially feasible to offer all attribute levels for one's own product or service, and it is not possible to control the attribute levels of competitors' offerings. It also is not possible to observe choices for alternatives that as yet do not exist. Even when the alternatives are available for choice, people may not be able to select their preferred one. For example, they may not be able to afford it, or it may not be available at the time when or place where a choice must be made. For these reasons, RP studies usually do not provide useful information to guide development of new products and services, and they often are ill suited for answering "what if" type questions about products that do exist.

Even if it is possible to conduct a field experiment following a well-designed experimental plan, RP field studies will be expensive. They will also be time consuming, which delays action based on the research. Field studies also are conducted "in the open." In competitive environments, they both tip off competitors and are susceptible to competitive sabotage.

Conjoint analysis (CA) and discrete choice experimentation (DCE), also known as choice-based conjoint analysis (CBC), are techniques developed since 1960 for avoiding the notable drawbacks of RP studies. The strategy with these approaches is to make inferences about the partworth of attribute levels from respondents' stated preferences (in CA) or stated choices (in DCE). Such studies are referred to as Stated Preference (SP) studies.

The preference or choice data collected in SP studies are expressed for abstract choice alternatives systematically constructed following an experimental design. The systematically constructed choice alternatives used in SP studies commonly are referred to as concept profiles. In marketing contexts, concept profiles typically describe brands, products, or services. In other areas of application, they may describe transportation, recreation, or health care options; public goods or policy choices; or any other sort of choice alternatives.

Concept profiles may consist of verbal descriptions, although they may include pen-and-ink representations, physical mockups, or videotaped demonstrations. The primary reason for restricting the choice situation in this way is to ensure that respondents evaluate each profile with respect to the same information. Ambiguous and equivocal cues are removed so that all respondents have at their disposal the same information and no more. When the concept describes an economic choice alternative, the description usually includes price.

The SP studies offer several advantages over RP studies. Compared to RP studies, SP studies may be conducted relatively rapidly and inexpensively. The use of abstract alternatives in the form of concept profiles reduces cost and execution time by providing prospective decision makers with what is thought to be the essential information that will ultimately drive preference or choice. Constructing concept profiles following appropriate experimental designs ensures that relevant data are available for estimating partworths of attribute levels. Since SP studies may be conducted in controlled settings, they may be conducted in a way that does not tip off competitors and is not susceptible to competitive sabotage. As with any controlled study, however, whether the results of the study may be generalized to real-life settings is open to the same generalizability issues that apply to all controlled experiments.

It turns out that the existing literature in the design of experiments contributes much to the problem of constructing concept profiles for SP studies. The design of experiments is a fascinating branch of statistics that is useful to all researchers in planning their investigations and drawing appropriate conclusions. Over the years, specialists in experimental design have developed a commonly used set of designs and models, and some of these are applicable to CA/DCE problems. The problem for applied researchers, however, is to modify commonly used designs for application to their specialized area and to develop new types of designs when necessary. Both approaches have occurred in the area of CA/DCE. The objectives of the present book are to overview and advance this work.

The purpose of this book is to discuss experimental design issues that arise in constructing concept profiles for DCE studies. Much of the material also applies to CA studies, however, as many issues that arise with DCE also arise when constructing concept profiles for CA studies.

The next two sections discuss CA and DCE, respectively. The emphasis with these techniques is on inferring respondents' valuation of concepts or the "components" of concept profiles. The objective of a typical study using these techniques is to determine the combination of components that would yield an alternative that would be most preferred by an individual or a group. The CA section expands on the logic of estimating partworths from stated preferences for concept profiles and provides an example of a CA study. It also provides some historical background that is relevant to both CA and DCE. The DCE section explains the difference between CA and DCE studies and provides an example of a study. It also discusses a number of issues that arise in the design of DCE studies. Chapters 3–8 of the book expand on these issues.

1.1 Conjoint Analysis

Simon (1957), Hoffman (1960, 1968), and Churchman (1961) were among the first to suggest that we could infer, or "capture," decision makers' policies and values by observing their decisions over enough circumstances. During the late 1960s and 1970s, psychologists working on a variety of seemingly unrelated problems developed a paradigm by which decision makers' policies might be inferred (Luce and Tukey, 1964; Krantz, 1964; Tversky, 1967; N. H. Anderson, 1981; Hoffman, Slovic, and Rorer, 1968). These researchers were primarily interested in determining the "composition rules" used by decision makers to combine information into overall judgments. One such group of researchers was interested in information integration, attitude change theory, person perception, and decision theory (N. H. Anderson, 1981). Another group of researchers was interested in how clinicians combined information to arrive at an overall diagnostic judgment (Dawes and Corrigan, 1974). The link to economics and the valuation of objects was provided by Lancaster (1966), who proposed that a consumer's utility for a good could be understood as a function of the utility for the components of the good (the partworths). That is, psychologists provided the view that the task of understanding how people choose could be understood in terms of how they combined information about the choice object, while Lancaster provided the insight that the relevant information to be combined was information about the components, or attributes, of goods.

The breakthrough insight in developing a paradigm for inferring partworths was that researchers may represent the composition rules decision makers use as main effects and interactions of an analysis of variance (ANOVA) model. Hoffman (1968) describes the rationale for the approach as follows:

> ANOVA models seem intuitively descriptive of many judgment situations, yet they have not previously been used to represent the judgment process. If judgment stimuli (cues) are regarded as categorical treatment factors rather than as random variables, and if the judgments made to the cues are considered as dependent variables then the inferential capabilities of the ANOVA technique can be applied to the study of judgment. The application is simple and direct: one prepares multidimensional judgmental stimuli by constructing all possible combinations (patterns) of the cue levels in a completely crossed factorial design. Such a set of patterns is of necessity orthogonal in the cue dimensions. (p. 340)

Let $y_{x_1x_2\ldots x_n}$ be the response or transformed response of the concept profile $(x_1, x_2, \ldots, x_n)$ of any choice set S_l. With a full factorial design, we model $y_{x_1x_2\ldots x_n}$ by

$$y_{x_1x_2\ldots x_n} = \mu + \sum_{i=1}^{n} \alpha_{x_i}^{A_i} + \sum_{\substack{i,j=1 \\ i\neq j}}^{n} \alpha_{x_ix_j}^{A_iA_j} + \sum_{\substack{i,j,k=1 \\ i\neq j\neq k\neq i}}^{n} \alpha_{x_ix_jx_k}^{A_iA_jA_k} + \ldots + \alpha_{x_1x_2\ldots x_n}^{A_1A_2\ldots A_n} + e_{x_1x_2\ldots x_n}, \quad (1.1)$$

where μ is the general mean; $\alpha_{x_i}^{A_i}$ is the effect of factor A_i at the x_i level; $\alpha_{x_ix_j}^{A_iA_j}$ is the effect of factors A_i and A_j at x_i and x_j levels, respectively; … , and $e_{x_1x_2\ldots x_n}$ is the iid (independently and identically distributed) random error. Under the diagnostic paradigm, composition rules used by decision makers are inferred by determining which terms of Equation 1.1 are statistically significant. For example, a significant two-way interaction implies that a decision maker considers levels of one factor when determining how much weight to assign to another factor in making an overall judgment.

Rather than infer the composition rule used to combine information, CA/DCE methodologies typically *assume* a composition rule and infer partworths as the effects of the corresponding ANOVA model. For example, a linear model that includes brand and attribute effects may be represented as

$$y_{jx_1x_2\ldots x_n} = \mu + \alpha_j^B + \sum_{i=1}^{n} \alpha_{x_i}^{A_i} + e_{x_1x_2\ldots x_n}, \quad (1.2)$$

where α_j^B is the attribute whose levels are the different brands. Assuming Equation 1.2 implies that the researcher believes two-way and higher-order interactions of the attributes are negligible. The error term of Equation 1.2 includes that of Equation 1.1 and variation due to possible misspecification.

In general, by assuming less than a full factorial model, a subset of all possible concept profiles is selected that allows estimation of all parameters of the reduced model that decision makers are assumed to use when combining partworths to arrive at preferences. For example, there are 1,200 profiles that may be constructed from all factorial combinations of the five automobile attributes identified in the preceding section: Brand(5) × Price(5) × Doors(2) × Transmission(4) × Engine(3) = 600. It is impractical, if not impossible, to have respondents rate each of 600 profiles. If it is assumed that two-way and higher-order interactions are negligible, however, the model of Equation 1.2 may be taken as the operating composition rule and a main-effect experimental plan used to construct profiles.

Table 3.7 of Chapter 3 provides a main-effects plan in 25 runs for estimating brands plus attributes study for estimating the partworths of automobile brands plus the four attributes mentioned in the preceding section. For example, following the plan of Table 3.7, the first two concept profiles presented to respondents are shown in Exhibit 1.1. Respondents would rate each of the remaining 23 profiles. The respondents' ratings of each profile comprise the dependent variable.

Brands or their attribute levels are traditionally coded as dummy variables using effects coding or orthogonal polynomials (see Chapters 2 and 4). Dummy-coded attribute levels are the independent variables. The parameter estimates are interpreted as the partworth, or part utility, of the respective brands or attribute levels. A key premise in interpreting parameters as indicating something akin to "utility" is that respondents must make trade-offs among attribute levels in arriving at their preference judgments. For example, the first two concept profiles require respondents to make trade-offs on all attributes. In commercial CA practice, parameters are estimated using ordinary least squares (OLS) regression.

Models that include interactions are also possible. Models that include interactions between brand and price might make sense in some situations. For example, in the case of automobiles, the price level that is considered expensive may depend on the brand; spending $41,000 on an economy brand may have much greater impact on preference than spending $41,000 on a luxury brand. Likewise, there may be interactions between brand or price and other product attributes. Leather upholstery may be expected with high-price or luxury brands, and accordingly its presence may contribute little to preference; it may not be expected with budget brands, and hence its presence might be considered a significant positive.

Green and Srinivasan (1978, 1990), Wittnik and Cattin (1982, 1989), and Green, Krieger, and Wind (2001) provide reviews of early CA research. A nonexhaustive list of studies of CA applications includes those by Green and Krieger (1991); Ryan, McIntosh, and Shackley (1998); Oppewal, Louviere,

EXHIBIT 1.1

Concept Profiles for Automobile Attributes

Concept Profile 1

Brand	Honda Accord
Price	$21,000
Doors	4 doors
Engine	8 cylinder
Transmission	Manual 6 speed

How likely is it that you would purchase the above automobile?

No Possibility	Very Unlikely	Somewhat Unlikely	Unsure	Likely	Very Likely	Certain
☐	☐	☐	☐	☐	☐	☐

Concept Profile 2

Brand	Ford Taurus
Price	$41,000
Doors	2 doors
Engine	8 cylinder
Transmission	Automatic 5 speed

How likely is it that you would purchase the above automobile?

No Possibility	Very Unlikely	Somewhat Unlikely	Unsure	Likely	Very Likely	Certain
☐	☐	☐	☐	☐	☐	☐

and Timmermans (1994); Darmon and Rouziès (1999); Katoshevski and Timmermans (2001); and Gustafsson, Herrmann, and Huber (2007).

An advantage of the ratings-based, CA approach is that respondents provide enough information so that parameters may be estimated at the individual level. These partworth estimates are often subsequently submitted to a clustering algorithm. The resulting clusters are interpreted as market "segments." Using partworth estimates in this way is called *post hoc* segmentation. Members of each segment are similar with respect to their preferences for product or service attribute levels.

Post hoc segments may be used in several ways, such as the following:

- The preferences of some segments may be more favorable to the existing brands of a firm than others, and marketers may focus effort on segments for which they have a differential advantage.
- Alternatively, they may identify segments that are attractive because of their size, rate of consumption, or lack of competition.

Information about utilities of the segments may then be used to design products or services that match what members of the segments desire.

- They may profile members of segments with demographic, geographic, or lifestyle variables and the like to gain insights on how to distribute or promote the product or service.

The focus of traditional CAs on direct elicitation of preference by using ratings has advantages and disadvantages. An important advantage is that ratings reduce responded burden by reducing the number of judgments required of respondents: When they are not replicated, the number of ratings required is equal to the number of alternatives. However, use of ratings scales has consequences. First, respondents use scales in different ways. Some use all the categories; some do not. Some distribute their responses at one end of the scale. Others favor the opposite end. In addition, predictions based on ratings are for preferences. A second, "choice simulator" stage must be introduced to link predicted preferences with choice proportions (Wiley and Low, 1983; Green and Krieger, 1988; Finkbeiner, 1988).

The traditional approach to CA also has an opportunity cost. In competitive markets, the success of a firm is influenced by both its own *and* its competitors' efforts. For example, a firm's market share is influenced not only by its own price but also by its competitors' prices, and the impact on market shares of a change in its own price depends on whether the change is matched by competitors. Traditional CA procedures are suited to capturing the effects of the efforts of the firms. They are ill suited for capturing the effects of competitors' actions because traditional conjoint studies do not require respondents to make trade-offs *between* profiles, only between *levels* of attributes for a single profile.

The potential shortcomings of rating methods have led marketers to look for alternative ways to capture responses. In recent years, attention has shifted to data collection and analysis procedures that are appropriate for *choice data*. DCE procedures do involve trade-offs between competing profiles.

1.2 Discrete Choice Experimentation

With DCE, the investigator organizes profiles into K systematically constructed choice sets of k_i profiles of m brands, $k_i \leq m$, where k_i is the number of profiles in set i, $i = 1, 2, \ldots, K$. In a choice experiment, n respondents are presented a series of such sets, and for each set they are asked to select the profile they consider to be the best. Work is only beginning on how to construct choice sets for various problems. Even at the current stage, however, DCE offers two advantages compared with traditional CA.

- The dependent variable is *choice*, which is similar conceptually to the act preceding purchase.
- With DCE, the respondent implicitly must consider the other alternatives in the choice set when making a choice. DCE therefore is suitable for detecting the effect that the *availability* of other alternatives has on attractiveness of a brand.

A nonexhaustive list of areas in which DCE has been used includes environmental science (Adamowicz, Boxall, and Williams, 1998; Bullock, Ellston, and Chalmers, 1998); geography (Oppenwal, Timmermans, and Louviere, 1997; van der Waerden, Oppewal, and Timmermans, 1993); health (Propper, 1995; Ryan and Farrar, 2000); marketing (Kamakura and Srivastava, 1984; Johnson and Olberts, 1991; Koelemeijer and Oppewal, 1999; Moore, Louviere, and Verma, 1999); tourism (Haider and Ewing, 1990); and transportation (Hensher, 1989; Fowkes and Wardman, 1988; Brandley and Gunn, 1990).

The design shown in Table 3.7 of Chapter 3 also may be used as the design for a DCE experiment. For example, the 25 runs of the design describe 25 profiles that may be organized into five choice sets of five profiles. A no-choice (or delayed-choice) option is added to each choice set. The first two sets of choice profiles presented to respondents based on this plan are shown in Exhibit 1.2.

Unlike traditional CA, with DCE respondents do not provide sufficient data so that it may be analyzed at the individual level. Respondents make a single selection from each choice set they are presented. The dependent variable consists of "shares," the proportion of N respondents selecting each alternative in a choice set. These proportions typically are modeled as the outcome of a multinomial process using the multinomial logit (MNL) random utility model (RUM; McFadden, 1974).

We make three points regarding the use of the MNL model before turning to choice set design issues.

- The model predicts that changes in shares are proportional to changes in preferences.
- Logits are heteroscedastic and correlated, so generalized least squares (GLS) or full information, maximum likelihood (FIML) algorithms should be used in estimation.
- As often is the case with nonlinear models, the covariance matrix of the observations depends on model parameters. This makes it difficult to derive global statements about optimal experimental designs.

EXHIBIT 1.2

Choice Sets for Automobile Profile

Suppose your current automobile needs to be replaced.

Choice Set 1

If the following automobiles were the only ones available, which one would you choose?

	Automobile 1	Automobile 2	Automobile 3	Automobile 4	Automobile 5
Brand	Honda Accord	Ford Taurus	Volvo C30	Ford Taurus	Ford Taurus
Price	$21,000	$41,000	$26,000	$31,000	$26,000
Doors	4 door	2 door	2 door	4 door	4 door
Engine	8 cylinder	8 cylinder	8 cylinder	6 cylinder	4 cylinder
Transmission	Manual 6 speed	Automatic 5 speed	Automatic 6 speed	Automatic 6 speed	Manual 5 speed
	☐	☐	☐	☐	☐

If the above automobiles were the only ones available, I would delay my purchase.

☐

Choice Set 2

If the following automobiles were the only ones available, which one would you choose?

	Automobile 1	Automobile 2	Automobile 3	Automobile 4	Automobile 5
Brand	Mitsubishi Lancer	Chevrolet Malibu	Volvo C30	Mitsubishi Lancer	Ford Taurus
Price	$36,000	$36,000	$41,000	$21,000	$36,000
Doors	4 door	4 door	4 door	2 door	2 door
Engine	4 cylinder	8 cylinder	6 cylinder	6 cylinder	6 cylinder
Transmission	Automatic 6 speed	Manual 5 speed	Manual 5 speed	Automatic 5 speed	Manual 6 speed
	☐	☐	☐	☐	☐

If the above automobiles were the only ones available, I would delay my purchase.

☐

1.3 Random Utility Models

Let u_i be the utility of a concept profile. The random utility model assumes that the utility may be partitioned into a systematic component v_i and a random component ε. Thus,

$$u_i = v_i + \varepsilon, \tag{1.3}$$

where $i = 1, \ldots, q \in S$, with q the number of alternatives in choice set S. The v_i components are assumed to be common to all people in the relevant population, while the ε components are assumed to be iid and to vary independently across individuals and choice sets.

The assumption of independence for the error terms is fairly restrictive, and a number of models have been developed to accommodate lack of independence. One such model is the availability/cross-effects model discussed in Chapter 6. The primary emphasis in this book, however, is on providing experimental designs appropriate for certain classes of problems that arise primarily in the context of DCE. That is, our emphasis is on providing designs that are appropriate for specific experimental situations rather than on modeling errors that deviate from the iid and other independence assumptions that characterize the most widely adopted estimation and hypothesis testing approaches used in DCE. If lack of independence is of specific interest or concern, the investigator may use alternative models, such as generalized extreme value (GEV), mixed logit, or probit. Alternatively, the independence issue may be addressed directly, for example, by presenting respondents with a single choice set (presumably at the cost of having to have more respondents). The designs discussed in this and following chapters are applicable regardless of the estimation and hypothesis testing procedures used.

Consider the choice set S with q elements having the utilities $u_1, u_2, \ldots, u_q$. The RUM assumes that on each choice occasion individuals choose the alternative they perceive to have the greatest utility on that occasion; that is, alternative i is chosen in preference to alternative j if and only if (iff)

$$u_i > u_j, \text{ for all } j \neq i \in S, \tag{1.4}$$

where $i \neq j = 1, \ldots, q \in S$. The observed choice proportions $p_1, p_2, \ldots, p_q$ for set S are the proportion of random draws (choices by n respondents) from a q-variate distribution of each variate (u_i, $i = 1, \ldots, q$) having the largest (extreme) value, where the "draws" are the stated choices of respondents.

There are two key points to be made regarding Equation 1.4. First, it is the *order* of the utilities that determines choice, not their absolute values. For example, rearranging terms, from Equations 1.3 and 1.4 we have alternative i chosen over alternative j iff

$$(v_i - v_j) > (\varepsilon_j - \varepsilon_i).$$

In general, alternative *i* will be selected iff

$$(v_i - v_j) > (\varepsilon_j - \varepsilon_i) \text{ and } (v_i - v_{j'}) > (\varepsilon_{j'} - \varepsilon_i), \text{ for all } i, j \neq j' \neq i \in S.$$

The probability that a given respondent chooses alternative *i* from choice set S is

$$P(\varepsilon_j - \varepsilon_i < v_i - v_j \ \forall j \neq i),$$

which is given by the cumulative $(q - 1)$ variate distribution that *each* term $\varepsilon_j - \varepsilon_i$ is below the observed value $v_i - v_j$.

With four alternatives, the covariance matrix **V** of three error differences $\varepsilon_2 - \varepsilon_1$, $\varepsilon_3 - \varepsilon_1$, and $\varepsilon_4 - \varepsilon_1$ is

$$V = \begin{pmatrix} \sigma_{11} + \sigma_{22} - 2\sigma_{12} & \sigma_{11} + \sigma_{23} - \sigma_{12} - \sigma_{13} & \sigma_{11} + \sigma_{24} - \sigma_{12} - \sigma_{14} \\ . & \sigma_{11} + \sigma_{33} - 2\sigma_{13} & \sigma_{11} + \sigma_{34} - \sigma_{13} - \sigma_{14} \\ . & . & \sigma_{11} + \sigma_{44} - 2\sigma_{14} \end{pmatrix}, \quad (1.5)$$

where σ_{ij} is the covariance of ε_i and ε_j. Equation 1.5 shows that the difference scores will be correlated even if the covariances of ε_j are zero. For example, the covariance term for the difference scores of alternatives 1 and 3 and 1 and 4 both contain the term σ_{11}.

1.4 The Logistic Model

The logistic model originally proposed by Luce (1959) is by far the most widely used model in DCE. It provides a closed-form expression for the cumulative distribution given by Equation 1.5 and an attractive theoretical underpinning. The cumulative distribution for the logistic distribution is the cumulative distribution for the differences of iid extreme value variates. McFadden (1974) proves the converse: The cumulative logit distribution for choice probabilities implies that the errors in utility values are extreme value. This distribution is also referred to as the *type 1 Gumbel extreme value distribution.*

The distribution function $F(x)$ for the extreme value variate is

$$F(x) = 1 - \exp\{-\exp[(x - a)/b]\}.$$

Random numbers of the extreme value variate ε: a, b can be computed from the relationship

$$x = a + b \ln\{\ln[1/(1 - \text{rndu}(x)]\}, \tag{1.6}$$

where the location parameter a is the mode, the scale parameter b is > 0, rndu(x) is a unit rectangular variate, and variance of x is $b^2\pi^2/6$. The mean of an extreme value variate is −0.57721. However, the mean is not relevant provided the distributions are iid as choice is driven by the *differences* between distributions.

Consider a choice set S with q choice alternatives having partworths u given by Equation 1.3. Let ε_i: a, b; $i = 1, \ldots, q$ be q extreme value variates having common scale parameter b and mode a. With the logit model, the population probability of choosing alternative i from choice set S may be written as

$$p_{i \in S} = e^{v_i} / \sum_{j \in S} e^{v_j}, \tag{1.7}$$

where v_i refers to the true utility for i, and $\sum_{j \in S} e^{v_j}$ refers to the sum over the alternatives in set S. It is evident that $\sum_{i \in S} p_i = 1.00$.

For purposes of estimation, a "base" alternative is added to each set. The base alternative is usually taken to be "no choice" or "delayed choice," in which case assigning it a systematic value v_{base} of zero has the logical interpretation of implying that the incremental value gained by not choosing is zero. Sometimes, base values other than no choice or delayed choice are used (e.g., the current product of the firm, a competitor's brand, the dominant brand in a market, or the present way of doing things). The reason for doing this may be to assess explicitly the impact of a change in the product, brand, or way of doing things on current market shares.

When a base option is added to each choice set, the result is $q + 1$ alternatives in each choice set. In this case, the probability that the ith alternative will be chosen from set S is a function of the differences $\varepsilon_i - \varepsilon_{\text{base}}$:

$$P\left([\varepsilon_j - \varepsilon_{\text{base}}] - [\varepsilon_i - \varepsilon_{\text{base}}] < [v_i - v_{\text{base}}] - [v_j - v_{\text{base}}], \forall j \neq i \right).$$

It is important to note that the estimated values for alternatives are actually contrasts, differences between the value of the alternative and the value of the base option. The probability of choosing alternative i with a base option is the probability that the *difference* ($v_i - v_{\text{base}}$) is greater than ($v_j - v_{\text{base}}$) for i, $j \neq i$. A change of base from no choice to (for example) the dominant brand in a market will affect the magnitudes and even the signs of utilities that are estimated as contrasts between the alternatives and a base.

The dependent variable with the logit model consists of $l = \ln(p_{ij}/p_{i,base})$, the log of the ratio of p_{ij} to $p_{i,base}$, where p_{ij} is the proportion of respondents picking the jth alternative in choice set i, and $p_{i,base}$ is the proportion picking base option in choice set i.

The log odds choice probabilities of each alternative i to the base alternative ($i = 0$) are

$$l_{i \in t_j} = \ln\left(\frac{p_{i \in t_j}}{p_{0 = base \in t_j}}\right) = \ln\left(\frac{e^{v_i} / \sum_{i \in S} e^{v_i}}{e^{v_0 = base} / \sum_{i \in S} e^{v_i}}\right) = v_i - v_0. \tag{1.8}$$

Equation 1.8 implies IIA (independence from irrelevant alternatives); that is, the addition of an alternative to the set will not affect the log odds ratios of pairwise choice proportions. An alternative interpretation is that changes in shares are proportional to changes in preferences. The IIA property is the focus of ongoing research because there is much evidence that it is violated in both individual and group choice data. In reality, alternatives do not compete equally with one another. Many examples of such effects have been documented (Becker, DeGroot, and Marschak, 1963; Bass, Pessemier, and Lehmann, 1972; Huber and Puto, 1983). Consequently, the choice of alternatives should be made with care when the generic form logit model is used so alternatives in choice sets that are likely to significantly violate its assumptions are not included in a choice set.

1.5 Contributions of the Book

Discrete choice experimentation introduces many challenging problems into the design of experiments. In many applications, the features of alternatives often—if not usually—are cost and benefits. That is, the "value" of attribute levels (from the perspective of the respondent) monotonically decreases or increases with increasing level of the attribute. In these situations, the profiles in a choice set should be designed so that obvious choices are ruled out, such as sets involving choices for which one alternative dominates, or is dominated, by other alternatives in the choice set. Sets in which no alternative dominates or is dominated are called *Pareto Optimal* choice sets.

Furthermore, when an alternative (whether a policy, transportation mode, or brand) is selected from a choice set, it is assumed that the choice is driven by the features of the alternative selected *and* the other available alternatives. That is, in choice situations, features of alternatives are compared with those of other available alternatives. The attractiveness of an alternative in the absence of other alternatives is referred to as the *own effect* of the alternative.

The effects of other alternatives available in a choice set on the attractiveness of the alternative are referred to as *cross effects*. Designs allowing the estimation of own and cross effects are called *availability designs*. Creating these sets presents interesting and challenging design problems.

The migration of survey research to the Internet poses several problems for CA/DCE. One of these problems is the limited "screen real estate" available for presentation of choice sets. There are several strategies for dealing with this problem. One is to use designs that have small choice set sizes. A second is to create designs using subsets of attributes, subsets of levels, or both. All these strategies pose experimental design issues. They also involve significant costs in the form of the number of choice sets that comprise complete designs.

One way to reduce the number of choice sets required to estimate model parameters is to engage in sequential experimentation. Sequential experimentation involves identifying the set of profiles required to fit the most parsimonious model that might characterize a choice process. If this set adequately fits the data, the experimental process stops. Otherwise, choices for a minimal additional set of profiles are collected, and the fit to a somewhat more complex model is evaluated. If this model adequately fits the data, the experimental process stops. Otherwise, additional choices for an additional set of profiles are collected. The experimental design task with sequential experimentation is the creation of such sequential sets of profiles.

Finally, there are two considerations when *explicit* economic costs are associated with attribute levels: first is total cost of the concept profile and second is the allocation of the total costs among the attributes. Both factors may influence people's preferences for profiles. These issues have not been separated in the DCE literature. Mixture and mixture–amount designs used in industrial experiments may be used to separate amount versus mixture effects in DCE studies. The aim of this book is to make contributions to the design of DCE studies.

2

Some Statistical Concepts

2.1 Principles of Experimental Design

To draw valid and useful conclusions from any experiment, proper planning of the study is crucial. This would include preparation of the instrument to collect data and decisions such as how the instrument will be administered to subjects, the number of subjects needed, and the type of analysis that will be conducted. These and other issues are considered in the statistical design of experiments.

As an example, suppose a market researcher wants to know if customers identify a product by the manufacturer's label or the quality of the product. For simplicity, assume the product is manufactured by only two companies. The researcher plans to show the product to customers in the study without any identifying labels to see if they can identify the manufacturer based solely on the quality of the product. Assume that the researcher has the resources to obtain three items from each company and let us consider the following four plans for conducting the study:

A. Present the six items to one customer in pairs, informing the customer that each pair consists of one item from each company
B. Present the six items in pairs as in A but to three customers of different demographics
C. Present the six items to one customer in random order, informing the customer that there are three items of each manufacturer among the six
D. Present the six items to one customer in random order as in C but without giving the customer any information

Assume that the customers cannot identify the two brands based solely on the quality of the product. That is, they simply guess to identify the labels. Suppose that in a given experiment all brands are correctly identified. Can we assume that this result is "better than chance," for which an outcome less likely than .05 is taken as the criterion? In plan A, the probability of correctly identifying one item in each pair will be 1/2; hence, the probability of correctly identifying all the items is $(1/2)^3 = 0.125$. Using a significance level of .05, we cannot conclude that the customer identifies the brands based on quality even if all six items are correctly identified. This is also true for plan B, although plan B has wider inductive basis since the conclusions are based on customers of different backgrounds, and choices are more likely to be independent. In plan C, the probability of the respondent correctly identifying all six items is $1/\binom{6}{3} = 0.05$. This is a borderline case, and if all items are correctly identified, we may conclude at the .05 significance level that the customer identifies the brands from their quality. In plan D, the probability of correctly identifying all the six items is $1/2^6 = 0.016$. In this case, the correct identification of all items overwhelmingly supports the research hypothesis at the .05 significance level.

In both plan A and plan B, the number of observations was too small to draw valid conclusions. If the experimenter decides to use plan B, five pairs of items can be presented to five customers so that the probability of correctly identifying all items is now $(1/2)^5 = 0.031$. If all items are correctly identified, we can conclude that customers identify the brands by the quality at the .05 significance level. This example illustrates the need to use an appropriate plan to collect data and the need for increased replications if the study has to be performed in a particular way. Note that to avoid systematic errors the items must be presented to customers in random order. In design-of-experiments terminology, this is called *randomization,* while the number of items used for each manufacturer is called *replication.* Note that for A and B to be equivalent the choices must be independent (e.g., earlier choices must not influence subsequent choices). If this is not the case, we have a repeated measures design. The design and analysis of repeated measures designs is an important subset in the field of experimental design.

The concepts of experimental design originated from agricultural experiments and slowly found their way into other branches of experimentation. Design terminology such as plots, blocks, and treatments should be interpreted based on the context of the experiment. A *plot* is an ultimate experimental response unit on which observations are collected. In social research, these are the subjects, although they may be dyads (such as customers and suppliers), stores, or geographic regions. Response units may be organized into blocks. A *block* is a collection of homogeneous experimental response units and can be interpreted as a group of response units with similar characteristics. Response units (or blocks) are assigned to treatment groups. *Treatments* are the external interventions applied to the elements of treatment groups to observe differences in response.

Randomization, replication, and local control are three essential features of any well-designed experiment. *Randomization* in the assignment of response units to treatments randomizes systematic differences between treatment groups and allows us to use random sampling theory as a criterion for judging whether observed differences between treatment groups are larger than would be expected to occur by chance. *Replication* allows us to control the magnitude of differences that would be expected to occur by chance. The variation of responses between treatments will be compared to the variation of responses within treatments, for which the within-group variation is an indication of the variation expected to occur by chance. More replications increase the precision of the study.

Local control is used in a broad sense to imply the methods that reduce sources of variation in observations other than the variability due to random errors. One form of local control is to conduct an experiment in a controlled environment that eliminates extraneous external sources of variation. Internal sources of variation, such as systematic differences in response units, may be controlled through design (e.g., by organizing units into homogeneous blocks) or statistically by performing what is known as covariance analysis. Efforts at local control inevitably involve trade-offs between internal and external validity of an experiment. *Internal validity* refers to whether observed differences between treatment groups may correctly be attributed to the treatment (rather than some unknown alternative explanation). *External validity* refers to whether a valid inference that differences in treatments were the source of differences in observations may be generalized to nonexperimental settings.

2.2 Experimental versus Treatment Design

The manner in which treatments are applied to subjects is called *experimental design*; the manner in which treatments are created is called *treatment design*. As discussed in Chapter 1, for example, treatments in conjoint analysis (CA) and discrete choice experimentation (DCE) are presented in the form of profiles consisting of stated levels on attributes. The primary reason for presenting treatments as abstract representations of choice alternatives is to ensure that respondents evaluate each profile with respect to the same information. Other examples of treatment designs include the "laid-off worker" experiment in which a claim for government assistance is judged based on race, gender, age, marital status, and work history (reliable or unreliable; see the work of Sniderman and Grob, 1996).

Well-known experimental designs include the completely randomized design (CRD), randomized block design (RBD), Latin square design (LSD), balanced incomplete block design (BIBD), crossover design (COD), partially balanced incomplete block design (PBIBD), and so on. Factorial experiments and fractional replications are known treatment designs. For further reading on these designs, see the work of Raghavarao (1983) and Montgomery (1991).

TABLE 2.1

Balanced Incomplete Block Design

Customer	Brands Ranked
1	A, B, D, G
2	B, C, E, A
3	C, D, F, B
4	D, E, G, C
5	E, F, A, D
6	F, G, B, E
7	G, A, C, F

To illustrate, suppose we are interested in knowing whether customers have an equal preference for seven brands of a product denoted by A, B, C, D, E, F, and G. We recruit seven customers, ask each to rank all seven brands on a scale from 1 to 7, and analyze the rank data. This design is known as the RBD. If we think that ranking seven brands may result in respondent fatigue, we may decide to limit the ranking task to four brands per customer. The experimental design of Table 2.1 can be used. This is an example of a BIBD and is formally introduced in the next section.

Suppose we want to know how car buyers make purchase decisions based on the price, gas mileage, and trunk capacity of a car. Price, gas mileage, and trunk capacity are known as factors (or attributes). Let us consider two levels of each factor as follows:

$$\text{Price: } p_0 = \$21{,}000;\ p_1 = \$26{,}000$$

$$\text{Gas mileage: } g_0 = 20 \text{ mpg};\ g_1 = 25 \text{ mpg}$$

$$\text{Trunk space: } t_0 = 15 \text{ cu ft};\ t_1 = 25 \text{ cu ft}$$

We can create eight profiles by combining these three factors: $\{p_0g_0t_0, p_0g_0t_1, p_0g_1t_0, p_0g_1t_1, p_1g_0t_0, p_1g_1t_0, p_1g_0t_1, p_1g_1t_1\}$. For example, $p_1g_0t_1$ represents a car that costs $26,000, has gas mileage of 20 mpg, and has a trunk capacity of 25 cu ft. We can ask buyers to rank all eight profiles or subsets of profiles that form an appropriate experimental design. Formation of such treatment designs and their statistical considerations are discussed in Sections 2.4–2.7.

In CA and DCE, we can use experimental designs to form treatment designs. Consider seven factors (a, b, c, d, e, f, g) and two levels for each factor denoted by the subscripts 0 and 1. Consider the situation of seven customers ranking four brands as in Table 2.1. Identify the letters in each row of the table with the higher level of the factor and the missing letters with the lower level. We get a treatment design with seven treatments as follows: $\{a_1b_1c_0d_1e_0f_0g_1, a_1b_1c_1d_0e_1f_0g_0, a_0b_1c_1d_1e_0f_1g_0, a_0b_0c_1d_1e_1f_0g_1, a_1b_0c_0d_1e_1f_1g_0, a_0b_1c_0d_0e_1f_1g_1, a_1b_0c_1d_0e_0f_1g_1\}$.

2.3 Balanced Incomplete Block Designs and 3-Designs

When the experimental units (or subjects) are heterogeneous, we can group them into "blocks" of homogeneous units. Units within a block are then homogeneous, while units between blocks are usually heterogeneous. If each treatment is applied once in every block, the resulting design is called an RBD. Of course, such a design is not feasible if the number of treatments is large, say 10 or more. In such a situation, an incomplete block design can be used. One such design, the BIBD, is an arrangement of v treatments in b sets of size k such that

1. every treatment occurs at most once in a set;
2. every treatment occurs in exactly r sets;
3. every pair of distinct treatments occurs together in λ sets.

Here, v, b, r, k, and λ are called the parameters of the BIBD, and they satisfy the following relations:

$$vr = bk, \; r(k - 1) = \lambda\,(v - 1), \; b \geq v.$$

A BIBD is said to be symmetrical if $v = b$. Any two distinct blocks of a symmetric BIBD will have λ treatments in common. The design in Table 2.1 is a symmetric BIBD with parameters $v = b = 7$, $r = k = 4$, and $\lambda = 2$. We can verify that every treatment (brand) is judged by four customers, and each pair of brands is judged by two customers. There is a vast body of literature available on these designs, and those interested are referred to the work of A. P. Street and Street (1987) or Raghavarao and Padgett (2005).

By interchanging the roles of treatments and blocks in any BIBD, we can construct its dual design. The dual design of a BIBD will have λ treatments in common between any two blocks and is also called a *linked block design*. A symmetric BIBD is clearly a linked block design. Table 2.2 gives a BIBD with parameters $v = 9$, $b = 12$, $r = 4$, $k = 3$, and $\lambda = 1$. Table 2.3 gives its dual or linked block design.

A design of v treatments arranged in b sets of size $k_1, k_2, \ldots, k_b$ where not all k_i are equal and that satisfies the three conditions of a BIBD is called a *pairwise balanced design*. An example of a pairwise balanced design with $v = 4$, $b = 10$, k_1 through $k_6 = 2$; k_7 through $k_{10} = 3$; $r = 6$, and $\lambda = 3$ is given in Table 2.4 (from the work of Raghavarao and Padgett, 2005).

A generalization of the BIBD is to allow every *triplet* of distinct treatments to occur equally often in the blocks. Such a design is called a *3-design*, which is defined as an arrangement of v treatments in b blocks of size k such that

1. every treatment occurs at most once in a block,
2. every treatment occurs in exactly r blocks,

TABLE 2.2

BIBD with $v = 9, b = 12, r = 4, k = 3$, and $\lambda = 1$

Block	Treatments
1	A, B, C
2	D, E, F
3	G, H, I
4	A, D, G
5	B, E, H
6	C, F, I
7	A, E, I
8	B, F, G
9	C, D, H
10	C, E, G
11	B, D, I
12	A, F, H

TABLE 2.3

Linked Block Design: Dual of the BIBD in Table 2.2

Block	Treatments
A	1, 4, 7, 12
B	1, 5, 8, 11
C	1, 6, 9, 10
D	2, 4, 9, 11
E	2, 5, 7, 10
F	2, 6, 8, 12
G	3, 4, 8, 10
H	3, 5, 9, 12
I	3, 6, 7, 11

TABLE 2.4

Pairwise Balanced Design

{A,B}, {A,C}, {A,D}, {B,C}, {B,D}, {C,D}, {A,B,C}, {A,B,D}, {A,C,D}, {B,C,D}

3. every pair of distinct treatments occurs together in λ_2 blocks, and
4. every triplet of distinct treatments occurs together in λ_3 blocks.

Here, v, b, r, k, λ_2, and λ_3 are called the parameters of a 3-design, and they satisfy the following relations:

$$vr = bk,\ r(k-1) = \lambda_2(v-1),\ \lambda_2(k-2) = \lambda_3(v-2),\ b \geq 2(v-1).$$

TABLE 2.5

A 3-Design with Parameters $v = 8$, $b = 14$, $r = 7$, $k = 4$, $\lambda_2 = 3$, and $\lambda_3 = 1$

{A,B,D,H}, {B,C,E,H}, {C,D,F,H}, {D,E,G,H}, {E,F,A,H}, {F,G,B,H}, {G,A,C,H}, {C,E,F,G}, {D,F,G,A}, {E,G,A,B}, {F,A,B,C}, {G,B,C,D}, {A,C,D,E}, {B,D,E,F}

An example of a 3-design with parameters $v = 8$, $b = 14$, $r = 7$, $k = 4$, $\lambda_2 = 3$, and $\lambda_3 = 1$ is given in Table 2.5 (from the work of Raghavarao and Padgett, 2005). A nice review article on 3-designs is that by Kreher (1996).

In the context of availability designs, the optimal designs turn out to be 3-designs with unequal block sizes. An unequal block size 3-design is an arrangement of v treatments in b blocks of size $k_1, k_2, \ldots, k_b$, satisfying the four conditions listed for 3-designs. These designs are discussed by Raghavarao and Zhou (1998). In the 3-design in Table 2.5, deleting the symbol H would result in an unequal block size 3-design with k_1 through $k_7 = 3$ and k_8 through $k_{14} = 4$.

2.4 Factorial Experiments

As discussed in Section 2.2, the manner in which treatments are applied to subjects is called experimental design, while the manner in which treatments are created is called treatment design. Factorial experiments are essentially treatment designs. Consider a study involving m factors (or attributes) with the ith factor at s_i levels ($i = 1, 2, \ldots, m$). Such an experiment is called an $s_1 \times s_2 \times \ldots \times s_m$ experiment. If $s_1 = s_2 = \ldots = s_m$ (= s, say), the factorial experiment is said to be symmetrical and is written as s^m. If at least two s_i are different, the experiment is said to be asymmetrical.

The advantage of factorial experiments is that we can study both the main effects and the interactions of factors. For simplicity, let us consider a car-purchasing situation in which the experimenter is studying two attributes, price and gas mileage, at two levels each:

$$\text{Price: } p_0 = \$21{,}000;\ p_1 = \$26{,}000.$$

$$\text{Gas mileage: } g_0 = 20 \text{ mpg};\ g_1 = 25 \text{ mpg}.$$

The experimenter creates four profiles by combining the two levels of each factor to get $\{p_0g_0, p_0g_1, p_1g_0, p_1g_1\}$. Each profile is shown to r customers, and they are asked to rate the profiles on a scale of 1–10 points where 1 indicates the least preferred and 10 indicates the most preferred. Let $T_{p_ig_j}$ be the total rating received from r customers by the profile p_ig_j ($i = 0, 1$ and $j = 0, 1$). Then, the change in rating when price is increased from \$21,000 to \$26,000 for cars with gas mileage $g_0 = 20$ mpg is

$$\frac{1}{r}\{T_{p_1g_0} - T_{p_0g_0}\}.$$

Similarly, the change in rating when price is increased from \$21,000 to \$26,000 for cars with gas mileage g_1 = 25 mpg is

$$\frac{1}{r}\{T_{p_1g_1} - T_{p_0g_1}\}.$$

The average of these two quantities gives us the overall change in rating when price is increased from \$21,000 to \$26,000 irrespective of the level of gas mileage. This is called the *main effect of price,* denoted by *P,* and is given by

$$P = \frac{1}{2r}\{T_{p_1g_0} - T_{p_0g_0} + T_{p_1g_1} - T_{p_0g_1}\}.$$

Similarly, the main effect of gas mileage, denoted by *G,* is given by

$$G = \frac{1}{2r}\{T_{p_0g_1} - T_{p_0g_0} + T_{p_1g_1} - T_{p_1g_0}\}.$$

It is possible that the change in rating when price is increased from \$21,000 to \$26,000 will not be the same at the two gas mileage levels. The difference in the change of rating when price is increased at 20 mpg and 25 mpg represents the interaction between price and gas mileage. It is denoted by *PG* and is given by

$$PG = \frac{1}{2r}\{(T_{p_1g_1} - T_{p_0g_1}) - (T_{p_1g_0} - T_{p_0g_0})\}.$$

Thus, by combining levels of price and gas mileage, the experimenter can estimate the main effects of price and gas mileage as well as how the attributes interact in the customer ratings. The interaction between two factors may exhibit synergism or antagonism. In this example, it is possible that a car with high price and better gas mileage may get much higher ratings (synergism) or much lower ratings (antagonism). Planning factorial experiments is the only way a researcher can get insight into the nature of interactions between attributes.

In the main effect *P,* the expression

$$T_{p_1g_0} - T_{p_0g_0} + T_{p_1g_1} - T_{p_0g_1}$$

is called a *contrast*. Note that the sums of coefficients of the contrasts are zero. In fact, the coefficients of the contrast for the interaction term *PG* can be obtained by multiplying the corresponding coefficients of the two main effects. Table 2.6 gives the plus and minus signs to be used for estimating

TABLE 2.6

Main Effects and Interaction in a 2^2 Experiment

	Factorial Effects			
Total Rating	I	P	G	PG
$T_{p_0g_0}$	+	–	–	+
$T_{p_1g_0}$	+	+	–	–
$T_{p_0g_1}$	+	–	+	–
$T_{p_1g_1}$	+	+	+	+

the main effects of price and gas mileage and their interaction. The columns represent the main effects P and G, their interaction term PG, and I, the total of the entire experiment. The rows represent the total rating received by each profile. For example, to find the contrast for estimating the main effect P, we multiply the signs in column P to the corresponding totals and add them to get $-T_{p_0g_0} + T_{p_1g_0} - T_{p_0g_1} + T_{p_1g_1}$, which is the same as the constant multiple of the main effect of P given here.

We now consider the general case of the factorial experiment. Consider a study involving m factors (or attributes) with the ith factor at s_i levels ($i = 1, 2, \ldots, m$). Let the s_i levels of factor i be represented by $\alpha_i = 0, 1, 2, \ldots, s_i - 1$. We can form $s_1 \times s_2 \times \ldots \times s_m$ profiles in all and obtain responses from subjects on each profile. Let $T_{\alpha_1\alpha_2\ldots\alpha_m}$ be the total of r responses on the profile with factor i at level α_i ($i = 1, 2, \ldots, m$). We can form $s_i - 1$ independent comparisons between the levels of the ith factor averaging over the remaining $m - 1$ factors, and these are

$$\frac{1}{(s_1s_2\ldots s_{i-1}s_{i+1}\ldots s_m)r}\left\{\sum T_{\alpha_1\alpha_2\ldots\alpha_{i-1}l\alpha_{i+1}\ldots\alpha_m} - \sum T_{\alpha_1\alpha_2\ldots\alpha_{i-1}0\alpha_{i+1}\ldots\alpha_m}\right\}, l = 1, 2, \ldots, s_i - 1.$$

In this expression, each of the $l = 1, 2, \ldots, s_i - 1$ summations includes $s_1\, s_2 \ldots s_{i-1}\, s_{i+1} \ldots s_m$ terms by assigning the values $0, 1, 2, \ldots, s_j - 1$ to α_j ($j \neq i$). In this sense, we say that the main effect of the ith factor has $s_i - 1$ degrees of freedom.

We can form $(s_i - 1)(s_j - 1)$ independent interaction terms between the levels of the ith and jth factors ($i < j$) by averaging over the remaining $m - 2$ factors, and these are

$$\frac{1}{(s_1s_2\ldots s_{i-1}s_{i+1}\ldots s_{j-1}s_{j+1}\ldots s_m)r}\left\{\sum T_{\alpha_1\alpha_2\ldots\alpha_{i-1}l\alpha_{i+1}\ldots\alpha_{j-1}l'\alpha_{j+1}\ldots\alpha_m}\right.$$
$$+\sum T_{\alpha_1\alpha_2\ldots\alpha_{i-1}0\alpha_{i+1}\ldots\alpha_{j-1}0\alpha_{j+1}\ldots\alpha_m} - \sum T_{\alpha_1\alpha_2\ldots\alpha_{i-1}l\alpha_{i+1}\ldots\alpha_{j-1}0\alpha_{j+1}\ldots\alpha_m}$$
$$\left.-\sum T_{\alpha_1\alpha_2\ldots\alpha_{i-1}0\alpha_{i+1}\ldots\alpha_{j-1}l'\alpha_{j+1}\ldots\alpha_m}\right\},$$

where $l = 1, 2, \ldots, s_i - 1$, and $l' = 1, 2, \ldots, s_j - 1$. In the preceding expression, each of the summations includes $s_1 s_2 \ldots s_{i-1} s_{i+1} \ldots s_{j-1} s_{j+1} \ldots s_m$ terms by assigning the values $0, 1, 2, \ldots, s_k - 1$ to α_k ($k \neq i, j$). The two-factor interaction of the ith and jth factors has $(s_i - 1)(s_j - 1)$ degrees of freedom. Similarly, by contrasting a two-factor interaction between the levels of a third factor, a three-factor interaction can be defined. A three-factor interaction between factors i, j, and k has $(s_i - 1)(s_j - 1)(s_k - 1)$ degrees of freedom. Extending this idea, we can define four and higher-order interaction terms as well.

Next, we discuss a 2^m experiment in more detail. Consider m factors $a_1, a_2, \ldots, a_m$ at two levels each and denote the two levels of factor a_i by a_{i0} and a_{i1}. There are 2^m profiles (also called runs or treatment combinations) of the form $a_{1j_1}a_{2j_2}\ldots a_{mj_m}$ where $j_i = 0$ or 1 for $i = 1, 2, \ldots, m$. Let r responses be taken on each profile and let $T_{a_{1j_1}a_{2j_2}\ldots a_{mj_m}}$ be the total of the r responses on the subscripted profile. Following the previous discussion in this section, we define the main effect A_i ($i = 1, 2, \ldots, m$) as

$$A_i = \frac{1}{2^{m-1}r}(a_{11}+a_{10})(a_{21}+a_{20})\ldots(a_{i-1,1}+a_{i-1,0})(a_{i1}-a_{i0})$$
$$(a_{i+1,1}+a_{i+1,0})\ldots(a_{m1}+a_{m0}),$$

where we multiply the terms in the parentheses on the right-hand side algebraically and replace $a_{1j_1}a_{2j_2}\ldots a_{mj_m}$ by $T_{a_{1j_1}a_{2j_2}\ldots a_{mj_m}}$.

We define the two-factor interaction A_iA_j ($i, j = 1, 2, \ldots, m$; $i < j$) as

$$A_iA_j = \frac{1}{2^{m-1}r}(a_{11}+a_{10})(a_{21}+a_{20})\ldots(a_{i-1,1}+a_{i-1,0})(a_{i1}-a_{i0})(a_{i+1,1}+a_{i+1,0})\ldots$$
$$\ldots(a_{j-1,1}+a_{j-1,0})(a_{j1}-a_{j0})(a_{j+1,1}+a_{j+1,0})\ldots(a_{m1}+a_{m0}),$$

with similar conventions as those used in the definition of A_i. We define the three-factor interaction $A_iA_jA_k$ ($i, j, k = 1, 2, \ldots, m$; $i < j < k$) as

$$A_iA_jA_k = \frac{1}{2^{m-1}r}(a_{11}+a_{10})(a_{21}+a_{20})\ldots(a_{i-1,1}+a_{i-1,0})(a_{i1}-a_{i0})(a_{i+1,1}+a_{i+1,0})\ldots$$
$$\ldots(a_{j-1,1}+a_{j-1,0})(a_{j1}-a_{j0})(a_{j+1,1}+a_{j+1,0})\ldots$$
$$\ldots(a_{k-1,1}+a_{k-1,0})(a_{k1}-a_{k0})(a_{k+1,1}+a_{k+1,0})\ldots(a_{m1}+a_{m0}),$$

with similar conventions as those used in the definition of A_i and A_iA_j. Finally, we have the m-factor interaction

$$A_1A_2\ldots A_m = \frac{1}{2^{m-1}r}(a_{11}-a_{10})(a_{21}-a_{20})\ldots(a_{m1}-a_{m0}).$$

It is interesting to note that the sum of the products of coefficients of $T_{a_{1j_1}a_{2j_2}\cdots a_{mj_m}}$ between any two main effects or interactions is zero, and statistically we say that any two main effects or interactions are uncorrelated (or orthogonal). There are m main effects, $\binom{m}{2}$ two-factor interactions, $\binom{m}{3}$ three-factor interactions, and so on and only one m-factor interaction. The total number of main effects and interactions is

$$m+\binom{m}{2}+\binom{m}{3}+\ldots+1=2^m-1.$$

2.5 Fractional Factorial Experiments

As the number of factors in a 2^m experiment increases, the number of profiles (or runs) in the experiment grows rapidly. For example, a 2^5 experiment has 32 profiles, while a 2^7 experiment has 128. For this reason, complete factorial designs are feasible only if the number of factors is relatively small. For a large number of factors, the experimenter may not have the resources to run the full factorial experiment and may choose to run only a subset or fraction of the design. Such a design is called a *fractional factorial design*. Fractional factorial designs give us information on the main effects and low-order interactions, provided we can assume that some of the higher-order interactions are unimportant and negligible. Fractional factorial designs are commonly used in screening experiments in which a large number of factors are under consideration and the experimenter is interested in identifying a small number of important factors. Once these important factors and interactions are identified, they can be studied in more detail in later experiments.

For example, in a study involving three attributes at two levels each, the full 2^3 experiment would require eight profiles. If experimenters have the resources to obtain responses on only four profiles, they may choose to run a one-half fraction of the full 2^3 design. The question, then, is which four profiles from the full factorial should they chose? Instead of picking a random sample of four runs from eight, we try to pick a fraction or subset that allows the estimation of the three main effects. Denote the three factors by a, b, and c and the two levels of each factor by the subscripts 0 and 1. Let $T_{a_ib_jc_k}$ be the total rating received from r customers by the profile $a_ib_jc_k$ $(i, j, k = 0, 1)$. Replacing $T_{a_ib_jc_k}$ by $a_ib_jc_k$, the full 2^3 experiment has plus and minus signs as shown in Table 2.7.

Suppose we select the four profiles $a_1b_0c_0$, $a_0b_1c_0$, $a_0b_0c_1$, and $a_1b_1c_1$. This design is called a 2^{3-1} fractional factorial because it is a half fraction of a full 2^3 factorial experiment and has $2^{3-1} = 4$ profiles. Note that for these four profiles the three-factor interaction ABC has all plus signs in its column in Table 2.8, that is, the same sign as column I (which represents the

TABLE 2.7

Main Effects and Interactions in a 2^3 Design

	Factorial Effects							
Profiles	*I*	*A*	*B*	*C*	*AB*	*AC*	*BC*	*ABC*
$a_0b_0c_0$	+	−	−	−	+	+	+	−
$a_1b_0c_0$	+	+	−	−	−	−	+	+
$a_0b_1c_0$	+	−	+	−	−	+	−	+
$a_1b_1c_0$	+	+	+	−	+	−	−	−
$a_0b_0c_1$	+	−	−	+	+	−	−	+
$a_1b_0c_1$	+	+	−	+	−	+	−	−
$a_0b_1c_1$	+	−	+	+	−	−	+	−
$a_1b_1c_1$	+	+	+	+	+	+	+	+

TABLE 2.8

A 2^{3-1} Fractional Factorial Plan

	Factorial Effects							
Profiles	*I*	*A*	*B*	*C*	*AB*	*AC*	*BC*	*ABC*
$a_1b_0c_0$	+	+	−	−	−	−	+	+
$a_0b_1c_0$	+	−	+	−	−	+	−	+
$a_0b_0c_1$	+	−	−	+	+	−	−	+
$a_1b_1c_1$	+	+	+	+	+	+	+	+

total of the entire experiment). For this reason, we say that the defining relation of the fraction is $I = ABC$. The defining relation for any fractional factorial consists of the main effects and interactions that have the *same sign* for the selected profiles. The 2^{3-1} fraction plus and minus signs are shown in Table 2.8.

Note that in Table 2.8 the signs in columns A and BC are identical, i.e., the contrasts used to estimate the main effect A and the two-factor interaction BC are identical. In fact, when we estimate the main effect A, we are in reality estimating $A + BC$, and we say that A is *aliased* with BC. Using a one-half fraction of the full factorial experiment results in a loss of information. Therefore, to estimate the main effects A, B, and C in the 2^{3-1} experiment, we must assume that the two-factor interactions are negligible. This assumption is required because in the 2^{3-1} fraction the main effects are aliased with the two-factor interactions. We can determine the alias structure of the 2^{3-1} experiment by examining its defining relation $I = ABC$. Multiplying both sides of the defining relation by the main effect A and noting from Table 2.7 that $A^2 = B^2 = C^2 = I$, we get

$$A = A^2BC = IBC = BC.$$

That is, the main effect A is aliased with the two-factor interaction BC. Similarly, the main effect B is aliased with AC, and the main effect of C is aliased with AB. If it is reasonable to assume that the two-factor interactions have negligible effect, the 2^{3-1} experiment can be used to estimate the main effects.

Fractional factorial designs can be classified based on their *resolution*. A design is said to be a resolution III design if the main effects are not aliased with each other but are aliased with two-factor interactions, and two-factor interactions may be aliased with other two-factor interactions. The 2^{3-1} experiment described is a resolution III design. A design is said to be resolution IV if the main effects are not aliased with each other *or* the two-factor interactions, but the two-factor interactions are aliased with each other. For example, consider a one-half fraction of the 2^4 design with the defining relation $I = ABCD$. This fraction consists of the $2^{4-1} = 8$ profiles from the full factorial for which the four-factor interaction $ABCD$ has all plus signs in its column. Multiplying both sides of the defining relation $I = ABCD$ by any main effect (say A), we get

$$A = A^2BCD = IBCD = BCD,$$

and by any two-factor interaction (say AB), we get

$$AB = A^2B^2CD = ICD = CD.$$

We can see that main effects are not aliased with each other *or* the two-factor interactions, but the two-factor interactions are aliased with each other. Finally, a resolution V design is one in which main effects are not aliased with each other or the two-factor interactions, and two-factor interactions are not aliased with each other but are aliased with three or higher-order interactions. The one-half fraction of the 2^5 design (or 2^{5-1} design) with the defining relation $I = ABCDE$ is an example of a resolution V design. In general, a design is said to be of resolution R if the minimum word length (number of factors in the interaction) in the defining relation is R. In a resolution R design, no p-factor effect is confounded with any other effect containing less than $R - p$ factors. In large experiments, it is often the case that only main effects and some low-order interactions are of real interest. Therefore, when using fractional factorial designs, one should choose the fraction with resolution such that main effects and interactions of interest are not aliased with other effects of interest.

As the number of factors under consideration increases, it may be necessary to use even a smaller fraction of the full factorial. For example, in a study involving five attributes at two levels each, the full 2^5 experiment would require 32 profiles. If experimenters have the resources to obtain responses on only 8 profiles, they may choose to run a one-quarter fraction of the 2^5 design, the 2^{5-2} design. Denote the five factors by a, b, c, d,

TABLE 2.9

A 2^{5-2} Fractional Factorial Design

Main Effects					
A	B	C	$D = AB$	$E = AC$	**Profiles**
−	−	−	+	+	$a_0b_0c_0d_1e_1$
+	−	−	−	−	$a_1b_0c_0d_0e_0$
−	+	−	−	+	$a_0b_1c_0d_0e_1$
+	+	−	+	−	$a_1b_1c_0d_1e_0$
−	−	+	+	−	$a_0b_0c_1d_1e_0$
+	−	+	−	+	$a_1b_0c_1d_0e_1$
−	+	+	−	−	$a_0b_1c_1d_0e_0$
+	+	+	+	+	$a_1b_1c_1d_1e_1$

and e and assume the experimenter chooses those eight runs for which the interactions ABD, ACE, and $BCDE$ have the same sign as column I. The defining relation of this fraction is $I = ABD = ACE = BCDE$. ABD and ACE are called the *generators* of this fraction, while $BCDE$ is the *generalized interaction* of ABD and ACE since $(ABD)(ACE) = A^2BCDE = BCDE$. In general, the defining relation of any one-quarter fraction of a full factorial will have two generators and their generalized interaction. Since the minimum word length in its defining relation is three, this 2^{5-2} design is a resolution III design.

This 2^{5-2} design can be obtained by first writing down a full 2^3 design and then adding two columns for D (= AB) and E (= AC) since D is aliased with AB and E is aliased with AC. The design is shown in Table 2.9. For more details on 2^k and 3^k and fractional factorial designs, refer to the work of Montgomery (1991) and Raktoe, Hedayat, and Federer (1981).

2.6 Hadamard Matrices and Orthogonal Arrays

Orthogonal arrays are useful in designing experiments, and the discussion here starts with a simple example. The following is an example of an orthogonal array with two levels (0 and 1) and *strength* 2:

$$\begin{matrix} 0 & 0 & 0 \\ 0 & 1 & 1 \\ 1 & 0 & 1 \\ 1 & 1 & 0 \end{matrix}$$

This is an orthogonal array of strength 2 because if we pick any *two* columns, every possible combination of the levels (00, 01, 10, and 11) occurs the same number of times (here exactly once) in the two columns. We call this array an $OA(4, 3, 2, 2)$ since it has four rows, three columns, two levels, and strength 2, respectively.

In general, an $n \times k$ array A with entries from a set of s symbols is said to be an orthogonal array with s levels, strength t, and index λ if every $n \times t$ subarray of A contains each t-tuple based on s symbols exactly λ times as a row. We denote such an array by $OA(n, k, s, t)$. Note that it is not necessary to mention λ in this notation because it can be determined by the equation $\lambda = n/s^t$. In the example, every possible combination of the levels occurs exactly once in the two columns, so $\lambda = 1$, and we say that the orthogonal array has *index unity*. The next example is an $OA(8, 4, 2, 3)$, that is, it has eight rows, four columns, two levels, and strength 3.

$$\begin{array}{cccc}
0 & 0 & 0 & 0 \\
0 & 0 & 1 & 1 \\
0 & 1 & 0 & 1 \\
0 & 1 & 1 & 0 \\
1 & 0 & 0 & 1 \\
1 & 0 & 1 & 0 \\
1 & 1 & 0 & 0 \\
1 & 1 & 1 & 1
\end{array}$$

It is easy to verify that $\lambda = n/s^t = 8/2^3 = 1$ in this example. That is, in any three columns of the orthogonal array, each of the eight possibilities (000, 001, 010, 011, 100, 101, 110, 111) occurs exactly once. The orthogonal arrays presented so far are arrays with only two levels (0 and 1). An example of an orthogonal array $OA(9, 4, 3, 2)$ with three levels (0, 1, and 2) is as follows:

$$\begin{array}{cccc}
0 & 0 & 0 & 0 \\
0 & 1 & 1 & 1 \\
0 & 2 & 2 & 2 \\
1 & 0 & 2 & 1 \\
1 & 1 & 0 & 2 \\
1 & 2 & 1 & 0 \\
2 & 0 & 1 & 2 \\
2 & 1 & 2 & 0 \\
2 & 2 & 0 & 1
\end{array}$$

For a detailed study of orthogonal arrays and related topics, those with mathematical or statistical backgrounds are referred to the work of Raghavarao (1971), Dey and Mukherjee (1999), and Hedayat, Sloane, and Stufken (1999).

Hadamard matrices, named after the famous French algebraist Jaques Hadamard, are closely related to orthogonal arrays. Hadamard matrices are square matrices of +1s and −1s whose rows are orthogonal. In general, a Hadamard matrix **H** of order n is an $n \times n$ square matrix with the following properties:

i Each element of **H** is +1 or −1.

ii $H'H = HH' = nI$.

where H′ is the transpose of H.

From property ii, it is clear that the columns are also orthogonal. The following are Hadamard matrices of order 2 and 4:

$$\begin{bmatrix} +1 & +1 \\ +1 & -1 \end{bmatrix} \qquad \begin{bmatrix} +1 & +1 & +1 & +1 \\ +1 & -1 & +1 & -1 \\ +1 & +1 & -1 & -1 \\ +1 & -1 & -1 & +1 \end{bmatrix}$$

It can be shown that if a Hadamard matrix exists, then n is 2 or a multiple of 4, that is, $n = 2$ or $n = 4t$. The converse, that if n is a multiple of 4 a Hadamard matrix must exist, is a conjecture. The smallest multiple of 4 for which no Hadamard matrix has so far been constructed is 668. Any matrix obtained by permuting the rows or columns of **H** or by multiplying its rows or columns by −1 is also a Hadamard matrix. All matrices obtained in this manner are said to be *isomorphic* or *equivalent* to **H**.

We can start with a Hadamard matrix and multiply its rows and columns by −1 as needed to make every element in the first row and first column equal to 1. Such a Hadamard matrix is said to be *normalized*. If every element in the first column is equal to 1, we say the matrix is *seminormalized*. By identifying all the columns except the first with $n - 1 = 4t - 1$ factors, we can see that a seminormal Hadamard matrix is a design matrix for an orthogonal main effects plan for $4t - 1$ factors (at two levels each) in $4t$ runs. That is, we get an orthogonal main effects plan of a 2^{4t-1} fractional factorial experiment in $4t$ runs. Fractional factorial designs obtained from Hadamard matrices are resolution III designs. They are also called *saturated* designs since no degree of freedom is left for estimating error. More details on Hadamard matrices can be obtained from the work of Raghavarao (1971) and Hedayat, Sloane, and Stufken (1999).

2.7 Foldover Designs

Recall that in a resolution III design the main effects are not aliased with each other, but they are aliased with two-factor interactions, and two-factor interactions may be aliased with other two-factor interactions. If we start with a resolution III fractional factorial and add to it a second fraction with the signs for *all* factors reversed, the combined design is called a *foldover* design. This process, called *folding over a design,* breaks the alias links between the main effects and two-factor interactions, giving us a resolution IV design. That is, we can use the combined design, or foldover design, to estimate the main effects clear of the two-factor interactions.

Consider the 2^{5-2} resolution III design described in Section 2.5. ABD and ACE are the generators of this fraction, and the complete defining relation of this fraction is $I = ABD = ACE = BCDE$. It is clear from the defining relation that all five main effects are aliased with the two-factor interactions. To separate the main effects and two-factor interactions, we can run a second fraction with the signs of all the factors reversed. The combined design is given in Table 2.10.

The defining relation of the foldover design can be obtained from the defining relation of its fractions. When we fold over a resolution III design,

TABLE 2.10

Resolution IV Foldover Design Obtained by Folding Over the 2^{5-2} Design

Main Effects					
A	B	C	$D = AB$	$E = AC$	**Profiles**
−	−	−	+	+	$a_0b_0c_0d_1e_1$
+	−	−	−	−	$a_1b_0c_0d_0e_0$
−	+	−	−	+	$a_0b_1c_0d_0e_1$
+	+	−	+	−	$a_1b_1c_0d_1e_0$
−	−	+	+	−	$a_0b_0c_1d_1e_0$
+	−	+	−	+	$a_1b_0c_1d_0e_1$
−	+	+	−	−	$a_0b_1c_1d_0e_0$
+	+	+	+	+	$a_1b_1c_1d_1e_1$
A	B	C	$D = -AB$	$E = -AC$	**Second Fraction**
+	+	+	−	−	$a_1b_1c_1d_0e_0$
−	+	+	+	+	$a_0b_1c_1d_1e_1$
+	−	+	+	−	$a_1b_0c_1d_1e_0$
−	−	+	−	+	$a_0b_0c_1d_0e_1$
+	+	−	−	+	$a_1b_1c_0d_0e_1$
−	+	−	+	−	$a_0b_1c_0d_1e_0$
+	−	−	+	+	$a_1b_0c_0d_1e_1$
−	−	−	−	−	$a_0b_0c_0d_0e_0$

we are in fact changing the signs of the generators that have an *odd* number of letters. In the example given, the defining relation of the second fraction is $I = -ABD = -ACE = BCDE$. Comparing the defining relations of the two fractions, we can see that they have one word with the same sign ($BCDE$) and two words with different signs (ABD and ACE). The combined design will have all words with the same sign as generators (here $BCDE$) plus all the independent even products of the words with different signs. By even products, we mean that the words are taken two at a time, four at a time, and so on. In this example, the even product of $(ABD)(ACE)$ is $BCDE$, so the complete defining relation of the foldover design is simply $I = BCDE$. It is clear from this defining relation that the combined design is a resolution IV design. The example is an illustration of the foldover technique for obtaining resolution IV designs; however, note that an even better design for five factors in 16 runs is the resolution V, 2^{5-1} design with the defining relation $I = ABCDE$.

Recall that fractional factorial designs derived from Hadamard matrices are resolution III designs. Box and Wilson (1951) first described the method of obtaining resolution IV designs by folding over resolution III designs derived from Hadamard matrices. It is also interesting to note that if we add to a fractional factorial design a second fraction with the signs of only a *single factor* reversed, we can break the alias links between the main effect of that factor and its two-factor interactions.

2.8 Mixture Experiments

An experiment in which the factors are the ingredients or constituents of a mixture is called a *mixture experiment*. In a mixture experiment, the response is a function of the proportions of each ingredient, with the amount of each ingredient usually measured by weight, volume, or the like. Some well-studied examples of mixture experiments are gasoline blends combining two or more gasoline stocks; cake or other food formulations combining ingredients such as baking powder, flour, sugar, and water; and concrete formed by mixing sand, water, and cement.

Consider a mixture formed by mixing together three ingredients and let x_1, x_2, and x_3 represent the proportions of each ingredient. A mixture with the proportions $x_1 = 0.2$, $x_2 = 0.3$, and $x_3 = 0.5$ is called a *complete* mixture because it is made up of all three ingredients. A mixture with the proportions $x_1 = 0.7$, $x_2 = 0.3$, and $x_3 = 0.0$ is called a *binary* mixture because it is made up of only two ingredients, while a mixture with the proportions $x_1 = 1.0$, $x_2 = 0.0$, and $x_3 = 0.0$ is called a *pure* or *single-component* mixture since it is made up of only one ingredient. Note that in all three examples, $x_1 + x_2 + x_3 = 1$. Because of this constraint, the levels of the factors in a mixture experiment cannot be chosen independently; that is, if we know that $x_1 = 0.2$ and $x_2 = 0.3$, then x_3

must be 0.5. To generalize, suppose a mixture consists of m components and let x_i represent the proportion of the ith component, then

$$x_i \geq 0, i = 1, 2, \ldots, m$$

and

$$\sum_{i}^{m} x_i = 1.$$

Since the levels of the factors in a mixture experiment are not independent, this makes them different from the experiments we have described so far in this chapter.

For a mixture with two components, the feasible factor space includes all values of x_1, x_2 for which $x_1 + x_2 = 1$. This constraint can be represented graphically by a line segment. For a mixture with three components, the feasible factor space is a triangle in which the vertices represent pure mixtures (only one component), and the edges represent binary mixtures (only two components). In general, the experimental region for an m component mixture is a simplex, that is, a regularly sided figure with m vertices and $m - 1$ dimensions, and all design points are at the vertices, on the edges or faces, or in the interior of the simplex. A set of points uniformly spaced on a simplex is called a *simplex lattice design* (See Cornell, 1990.). Consider a mixture experiment with m components and let each component take the proportions given by the $q + 1$ equally spaced values from 0 to 1,

$$x_i = 0, 1/q, 2/q, \ldots, 1 \quad i = 1, 2, \ldots, m.$$

If all possible combinations of the proportions from this equation are used, the design is called an $\{m, q\}$ simplex lattice design for m components. For example, if $m = 3$ and $q = 2$, then

$$x_i = 0, 1/2, 1 \quad i = 1, 2, 3$$

and the {3, 2} simplex lattice design consists of the points

$$(1, 0, 0), (0, 1, 0), (0, 0, 1), (1/2, 1/2, 0), (1/2, 0, 1/2), (0, 1/2, 1/2).$$

Similarly, a {3, 3} simplex lattice design consists of the points

$$(1, 0, 0), (0, 1, 0), (0, 0, 1), (2/3, 1/3, 0), (1/3, 2/3, 0), (2/3, 0, 1/3), (1/3, 0, 2/3), (0,2/3, 1/3), (0, 1/3, 2/3), (1/3, 1/3, 1/3).$$

Those with a strong mathematical or statistical background are referred to the work of Cornell (1990) for a detailed study of this topic.

A first-order model is given by

$$E(y) = \beta_0 + \sum_{i=1}^{m} \beta_i x_i$$

where $E(y)$ is the expected value of y.

However, due to the fact that $x_1 + x_2 + \ldots + x_m = 1$, the parameters β_i in this equation are not unique. If we multiply the term β_0 in the equation by $x_1 + x_2 + \ldots + x_m = 1$, we get

$$E(y) = \beta_0(x_1 + x_2 + \ldots + x_m) + \sum_{i=1}^{m} \beta_i x_i$$

$$= \sum_{i=1}^{m} \beta_i^* x_i,$$

where $\beta_i^* = \beta_0 + \beta_1$. This equation is called the *canonical form* of the first-order mixture model. Removing the asterisk sign from the model, the canonical form of the linear model is

$$E(y) = \sum_{i=1}^{m} \beta_i x_i.$$

The general second-degree polynomial in m variables is

$$E(y) = \beta_0 + \sum_{i=1}^{m} \beta_i x_i + \sum_{i=1}^{m} \beta_{ii} x_i^2 + \sum_{i<j}\sum^{m} \beta_{ij} x_i x_j.$$

Again, applying the fact that $x_1 + x_2 + \ldots + x_m = 1$, and

$$x_i^2 = x_i\left(1 - \sum_{\substack{j=1 \\ j \neq i}}^{m} x_j\right),$$

we get

$$E(y) = \beta_0\left(\sum_{i=1}^{m} x_i\right) + \sum_{i=1}^{m} \beta_i x_i + \sum_{i=1}^{m} \beta_{ii} x_i\left(1 - \sum_{j \neq i}^{m} x_j\right) + \sum_{i<j}\sum^{m} \beta_{ij} x_i x_j$$

$$= \sum_{i=1}^{m} (\beta_0 + \beta_i + \beta_{ii}) x_i - \sum_{i=1}^{m} \beta_{ii} x_i \sum_{j \neq i}^{m} x_j + \sum_{i<j}\sum^{m} \beta_{ij} x_i x_j$$

$$= \sum_{i=1}^{m} \beta_i^* x_i + \sum_{i<j}\sum^{m} \beta_{ij}^* x_i x_j,$$

where the parameters in the last equation are simple functions of the original parameters, that is, $\beta_i^* = \beta_0 + \beta_i + \beta_{ii}$ and $\beta_{ij}^* = \beta_{ij} - \beta_{ii} - \beta_{jj}$, $i, j = 1, 2, \ldots, m$, $i < j$. Removing the asterisk sign from the model, the canonical form of the quadratic model is

$$E(y) = \sum_{i=1}^{m} \beta_i x_i + \sum_{i<j}\sum^{m} \beta_{ij} x_i x_j.$$

The βs used in these models are population parameters, and they are estimated from experimental data using methods discussed in Section 2.9.2.

2.9 Estimation

The estimation problem involves estimating the unknown parameters of known probability distributions of random variables. Commonly used models for probability distributions are functions of unknown parameters. We need to estimate these unknown parameters from the data and test hypotheses we may formulate on them. We discuss estimation in this section. Hypothesis testing is discussed in Section 2.11. Two commonly used estimation procedures are the least squares method and the maximum likelihood method. We first introduce some commonly used probability models and then discuss the estimation methods.

2.9.1 Probability Models

A random variable with only two outcomes is said to have a *Bernoulli distribution*. Let the two outcomes be called "success" (denoted by 1) and "failure" (denoted by 0). Let the probability of success, $P(X = 1) = p$, and the probability of failure, $P(X = 0) = 1 - p = q$. Then

$$E(X) = p \text{ and } Var(X) = pq.$$

Assume we have n trials, each with only two outcomes, such that (a) the outcome of one trial does not influence the outcome of any other trial, and (b) the probability of success p is the same for each trial. Let X represent the number of successes in n such trials, then X is said to have a binomial distribution with

$$P(X = x) = \binom{n}{x} p^x q^{n-x},$$

where $x = 0, 1, \ldots, n$. Here, $E(X) = np$ and $\mathrm{Var}(X) = npq$.

Assume we have n trials, each with $k + 1$ outcomes denoted by $E_0, E_1, \ldots, E_k$, such that (a) the outcome of one trial does not influence the outcome of any other trial, and (b) in each trial the outcome E_i occurs with probability p_i, $i = 0, 1, \ldots, k$, such that $\sum_{i=0}^{k} p_i = 1$. The number of trials X_i, in which E_i occurs ($i = 0, 1, \ldots, k$) is said to have a multinomial distribution with

$$P(X_i = x_i, i = 0, 1, \ldots, k) = \frac{n!}{\prod_{i=0}^{k} x_i!} \prod_{i=0}^{k} p_i^{x_i}, \qquad \text{if } \sum_{i=0}^{k} x_i = n$$

$$= 0 \text{ otherwise.}$$

We discuss the multinomial distribution and its transformation in Section 2.10.

A random variable X that can take any real value is said to have a normal distribution with mean μ and variance σ^2 if

$$P(x < X \le x + dx) = \frac{1}{\sqrt{2\pi}\sigma} e^{\frac{-(x-\mu)^2}{2\sigma^2}} dx.$$

Here, $1/\sqrt{2\pi}\sigma e^{-(x-\mu)^2/2\sigma^2}$ is called the *density function* of the random variable. We denote such a distribution by $N(\mu, \sigma^2)$.

2.9.2 Linear Estimation: Least Squares and Weighted Least Squares

In this section, we consider the expected value of the responses or suitably transformed responses to be linear functions of unknown parameters with known coefficients and estimate the parameters as linear functions of responses or transformed responses. Let there be n responses or transformed responses denoted by $y_1, y_2, \ldots, y_n$, and we assume

$$E(y_i) = \sum_{j=1}^{k} x_{ij}\beta_j,$$

where $\beta_1, \beta_2, \ldots, \beta_k$ are unknown parameters, and x_{ij} are known constants. Usually, x_{ij} are the demographic variables of the respondents or the characteristics of the items presented to the respondents. Let $\mathbf{X} = (x_{ij})$, $\mathbf{y}' = (y_1, y_2, \ldots, y_n)$, and $\beta' = (\beta_1, \beta_2, \ldots, \beta_p)$. We then have

$$E(y) = \mathrm{X}\beta.$$

Here, $\mathbf{X}$ is called the design matrix and is of order $n \times k$.

A linear parametric function $l'\beta$ is said to be estimable if there exists a vector **a** such that

$$E(\mathbf{a}'\,\mathbf{y}) = l'\beta.$$

It is interesting to note that not all linear functions $l'\beta$ are estimable with any design matrix **X**, and $l'\beta$ is estimable if and only if l is a linear combination of the rows of **X**. Here, **a′y** is called an *unbiased estimator* of $l'\beta$, and there may exist several unbiased estimators of $l'\beta$. Of the unbiased estimators, we like to choose the estimator with the smallest variance, known as the best linear unbiased estimator (BLUE) of $l'\beta$. In design literature, it is proved (see Chapter 1 in Raghavarao and Padgett's 2005 work) that the BLUE of $l'\beta$ is $l'\hat{\beta}$, where $\hat{\beta}$ is a solution of the equation

$$X'\Sigma^{-1}X\,\hat{\beta} = X'\Sigma^{-1}\mathbf{y},$$

Σ being the dispersion matrix of **y**. The equation is called a *weighted normal equation* and is obtained by minimizing $(\mathbf{y} - \mathbf{X}\,\beta)'\,\Sigma^{-1}(\mathbf{y} - \mathbf{X}\,\beta)$ with respect to β. Since $\hat{\beta}$ is obtained from weighted least squares, it is called the *weighted least squares estimator.* When $\Sigma = \sigma^2 I_n$, where $\mathbf{I}_n$ is the identity matrix, $\hat{\beta}$ is called the ordinary least squares (OLS) estimator.

When the interest is in estimating each of the individual k parameters β_1, β_2, … , β_k, we need the design matrix **X** such that $\mathbf{X}'\Sigma^{-1}\,\mathbf{X}$ is nonsingular. In that case,

$$\hat{\beta} = (X'\Sigma^{-1}X)^{-1}X'\Sigma^{-1}\mathbf{y}$$

are the estimated parameters with

$$\text{Var}(\hat{\beta}) = (X'\Sigma^{-1}X)^{-1}.$$

When $\Sigma = \sigma^2 I_n$, the least squares estimator of β is $(\mathbf{X}'\mathbf{X})^{-1}\mathbf{X}'\mathbf{y}$ with dispersion matrix $\sigma^2(\mathbf{X}'\mathbf{X})^{-1}$. We use these results in the example that follows.

Let us show a new product to four respondents and ask them to rate it on a 10-point scale. The demographics of the respondents and ratings are given as follows:

Respondent	Gender	Age	Rating
1	Male	< 30 years	9
2	Female	≥ 30 years	7
3	Female	< 30 years	8
4	Male	≥ 30 years	6

Here, $\mathbf{y}' = (9,7,8,6)$. There is expected to be a common mean β_1 denoted by μ for all the respondents. To get the **X** matrix, we code as follows: gender: male (0), female (1); age: < 30 years (1), ≥ 30 years (0).

$$X = \begin{bmatrix} 1 & 0 & 1 \\ 1 & 1 & 0 \\ 1 & 1 & 1 \\ 1 & 0 & 0 \end{bmatrix}$$

Assuming $\Sigma = \sigma^2 I_4$, we get

$$\begin{bmatrix} \hat{\mu} \\ \hat{\beta}_2 \\ \hat{\beta}_3 \end{bmatrix} = \begin{bmatrix} 4 & 2 & 2 \\ 2 & 2 & 1 \\ 2 & 1 & 2 \end{bmatrix}^{-1} \begin{bmatrix} 30 \\ 15 \\ 17 \end{bmatrix} = \begin{bmatrix} 6.5 \\ 0 \\ 2 \end{bmatrix}$$

$$\text{Var}\begin{bmatrix} \hat{\mu} \\ \hat{\beta}_2 \\ \hat{\beta}_3 \end{bmatrix} = \sigma^2 \begin{bmatrix} 3/4 & -1/2 & -1/2 \\ -1/2 & 1 & 0 \\ -1/2 & 0 & 1 \end{bmatrix}$$

$\hat{\beta}_3 = 2$ is the estimated difference in ratings for the respondents < 30 years and ≥ 30 years. In this example, $\hat{\beta}_2 = 0$, implying that the estimated difference in ratings between the genders is zero.

When σ^2 is unknown, it is estimated by

$$\hat{\sigma}^2 = (\mathbf{y} - X\hat{\beta})'(\mathbf{y} - X\hat{\beta})/(n-r) = \left[\mathbf{y}'\mathbf{y} - \hat{\beta}'X'\mathbf{y}\right]/(n-r),$$

where r is the rank of the design matrix **X**. In the illustration considered, **X** is of rank 3 and

$$\hat{\sigma}^2 = [230 - 229]/(4-3) = 1.$$

2.9.3 Maximum Likelihood Estimation

If $X_1, X_2, \ldots, X_n$ is a random sample, the joint probability of obtaining the sample values expressed as a function of unknown parameters is called the *likelihood*. Maximizing the likelihood with respect to the unknown parameters, we get maximum likelihood estimators of the parameters. Maximizing the likelihood is the same as maximizing the log likelihood.

Consider the multinomial normal distribution and suppose we observe x_0, $x_1 \ldots x_k$ frequencies. Then, the likelihood is

$$L(p_0, p_1, \ldots, p_k) = \frac{n!}{\prod_{i=0}^{k} x_i!} \prod_{i=0}^{k} p_i^{x_i}$$

and the log likelihood is

$$l(p_0, p_1, \ldots, p_k) = \ln L(p_0, p_1, \ldots, p_k) = \ln n! - \sum_{i=0}^{k} \ln x_i! + \sum_{i=0}^{k} x_i \ln p_i.$$

Differentiating $l(p_0, p_1, \ldots, p_k)$ with respect to p_i, for $i = 1, 2, \ldots, k$, noting $\sum_{i=0}^{k} p_i = 1$ and equating to zero, we get

$$-\frac{x_0}{1 - \sum_{i=1}^{k} \hat{p}_i} + \frac{x_i}{\hat{p}_i} = 0, \qquad i = 1, 2, \ldots, k.$$

This implies

$$\frac{x_1}{\hat{p}_1} = \frac{x_2}{\hat{p}_2} = \ldots = \frac{x_k}{\hat{p}_k} = \frac{x_0}{1 - \sum_{i=1}^{k} \hat{p}_i} = \frac{n}{1},$$

and hence $\hat{p}_i = x_i/n$, for $i = 0, 1, 2, \ldots, k$.

Let $X_1, X_2, \ldots, X_n$ be a random sample from $N(\mu, \sigma^2)$ where μ and σ^2 are unknown. Then, the likelihood is

$$L(\mu, \sigma^2) = \frac{1}{(2\pi\sigma^2)^{n/2}} e^{-\sum_{i=1}^{n} \frac{(x_i - \mu)^2}{2\sigma^2}};$$

and log likelihood is

$$l(\mu, \sigma^2) = \ln L(\mu, \sigma^2) = -\frac{n}{2} \ln 2\pi - \frac{n}{2} \ln \sigma^2 - \frac{\sum_{i=1}^{n} (x_i - \mu)^2}{2\sigma^2}.$$

Now, $\partial l(\mu, \sigma^2)/\partial\mu = 0$ gives $\hat{\mu} = \bar{x}$, while $\partial l(\mu, \sigma^2)/\partial\sigma^2 = 0$ gives $\hat{\sigma}^2 = \sum_{i=1}^{n} (x_i - \bar{x})^2/n$.

2.10 Transformations of the Multinomial Distribution

Suppose n respondents are asked to select the best of k items presented to them. The respondents are also presented with a "no-choice" option if they do not like any of the k choices presented. Let X_i denote the random variable indicating the number of people who select the ith item, $i = 1, 2, 3, \dots, k$, and let X_0 denote the random variable indicating the number of people who select none of the k items, that is, who select the no-choice option. Let p_i be the proportion of people selecting the ith item ($i = 1, 2, \dots, k$) on a long run, and let $p_0 = 1 - \sum_{i=1}^{k} p_i$. In other words, p_i denotes the probability that an individual selects the ith item for $i = 1, 2, \dots, k$, and p_0 denotes the probability that an individual selects the no-choice option. We assume that $p_i > 0$ for $i = 0, 1, \dots, k$.

The probability distribution of the $k + 1$ variables X_i for $i = 0, 1, 2, \dots, k$ is the *multinomial distribution* with

$$E(X_i) = np_i, \ \mathrm{Var}(X_i) = np_i(1 - p_i), \text{ and } \mathrm{Covar}(X_i, X_j) = -np_ip_j, \ i, j = 0, 1, \dots, k, \ i \neq j.$$

$E(X_i)$ and $\mathrm{Var}(X_i)$ denote the expected value and variance of X_i, respectively, and $\mathrm{Covar}(X_i, X_j)$ is the covariance of X_i and X_j. Using matrix notation and writing $\mathbf{X}' = (X_0, X_1, \dots, X_k)$, and $\mathbf{p}' = (p_0, p_1, \dots, p_k)$, we have

$$E(\boldsymbol{X}) = n\boldsymbol{p}, \text{ and } \mathrm{Var}(\boldsymbol{X}) = n\{D(\boldsymbol{p}) - \boldsymbol{p}\boldsymbol{p}'\},$$

where $E(\mathbf{X})$ and $\mathrm{Var}(\mathbf{X})$ are the expected value and dispersion matrix of the random vector $\mathbf{X}$, respectively; $D(\mathbf{p})$ is a diagonal matrix with diagonal elements indicated by the vector $\mathbf{p}$; and $\mathbf{p}'$ is the transpose of the vector $\mathbf{p}$. Here, $\mathrm{Var}(\boldsymbol{X})$ is singular; therefore, a G-inverse is given by

$$\{\mathrm{Var}(\boldsymbol{X})\}^{-} = \frac{1}{n}\{D(\boldsymbol{p})\}^{-1}.$$

If we want to work with nonsingular matrices, we can delete X_0 in $\boldsymbol{X}$ and p_0 in $\boldsymbol{p}$ to get the vectors $\mathbf{X}^*$ and $\mathbf{p}^*$ satisfying

$$E(\boldsymbol{X}^*) = n\,\boldsymbol{p}^*, \text{ and } \mathrm{Var}(\boldsymbol{X}^*) = n\{D(\boldsymbol{p}^*) - \boldsymbol{p}^*\boldsymbol{p}^{*\prime}\},$$

where $\mathrm{Var}(\boldsymbol{X}^*)$ is nonsingular.

The maximum likelihood estimator of p_i is $\hat{p}_i = X_i/n$ for $i = 0, 1, \dots, k$ as given in subsection 2.9.3. Customarily, transformed $\hat{p}_i$ are modeled using the demographics of the respondents and the characteristics of the items, as will be seen in further chapters. We now briefly introduce some useful transformations.

In DCE literature, the following logit transformation is used:

$$\hat{L}_{i0} = \ln\left\{\frac{\hat{p}_i}{\hat{p}_0}\right\} = \ln\left\{\frac{X_i}{X_0}\right\}.$$

By the delta method, we have

$E(\hat{L}_{i0}) = L_{i0}$ $\text{var}(\hat{L}_{i0}) = 1/n\ [1/p_i + 1/p_0]$ $\text{covar}\left(\hat{L}_{i0}, \hat{L}_{j0}\right) = 1/np_0;\ i, j = 1,$ $2, \ldots, k, i \neq j$, where $L_{i0} = \ln\{p_i/p_0\}$. Defining $\hat{\boldsymbol{L}}_0' = \left(\hat{L}_{10}, \hat{L}_{20}, \ldots, \hat{L}_{k0}\right)$ and $\boldsymbol{L}_0' = \left(L_{10}, L_{20}, \ldots, L_{k0}\right)$, we have

$$\mathrm{E}\left(\hat{\boldsymbol{L}}_0\right) = \boldsymbol{L}_0$$

and

$$\text{var}\left(\hat{\boldsymbol{L}}_0\right) = \frac{1}{n}\begin{pmatrix} \dfrac{1}{p_0} + \dfrac{1}{p_1} & \dfrac{1}{p_0} & \cdots & \dfrac{1}{p_0} \\ \dfrac{1}{p_0} & \dfrac{1}{p_0} + \dfrac{1}{p_2} & \cdots & \dfrac{1}{p_0} \\ \vdots & \vdots & \cdots & \vdots \\ \dfrac{1}{p_0} & \dfrac{1}{p_0} & \cdots & \dfrac{1}{p_0} + \dfrac{1}{p_k} \end{pmatrix}.$$

In the standard literature, however, the following transformation is considered the logit transformation of $\hat{p}_i$:

$$\hat{L}_i = \ln\left\{\frac{\hat{p}_i}{(1 - \hat{p}_i)}\right\} = \ln\left\{\frac{X_i}{(n - X_i)}\right\}.$$

By the delta method, we have

$E(\hat{L}_i) = L_i$, $\text{var}(\hat{L}_i) = 1/np_i(1 - p_i)$ $\text{covar}\left(\hat{L}_i, \hat{L}_j\right) = -1/n(1 - p_i)(1 - p_j)$ $i, j = 0,$ $1, \ldots, k, i \neq j$, where $L_i = \ln\{p_i/(1 - p_i)\}$. Defining $\hat{\boldsymbol{L}}' = \left(\hat{L}_0, \hat{L}_1, \ldots, \hat{L}_k\right)$ and $\boldsymbol{L}' = \left(L_0, L_1, \ldots, L_k\right)$ we have

$$\mathrm{E}(\hat{\mathbf{L}}) = \mathbf{L},$$

and

$$\operatorname{var}(\hat{\mathbf{L}}) = \frac{1}{n}\begin{bmatrix} \frac{1}{p_0(1-p_0)} & \frac{-1}{(1-p_0)(1-p_1)} & \cdots & \frac{-1}{(1-p_0)(1-p_k)} \\ \frac{-1}{(1-p_1)(1-p_0)} & \frac{1}{p_1(1-p_1)} & \cdots & \frac{-1}{(1-p_1)(1-p_k)} \\ \vdots & \vdots & \cdots & \vdots \\ \frac{-1}{(1-p_k)(1-p_0)} & \frac{-1}{(1-p_k)(1-p_1)} & \cdots & \frac{1}{p_k(1-p_k)} \end{bmatrix}$$

An alternative transformation used is the probit transformation. Let $\phi(z)$ be the density function of the standard normal variable given by

$$\Phi(z) = \frac{1}{\sqrt{2\pi}} e^{-z^2/2},$$

and $\Phi(z)$ be the cumulative distribution function of the standard normal variable given by

$$\Phi(z) = \int_{-\infty}^{z} \phi(z)\,dx.$$

Both z and $\phi(z)$ are given in normal probability tables in several statistical textbooks, where $\phi(z)$ are given in the main body of the table, bordered by z values. The probit transformation of $\hat{p}_i$ is

$$P_i(\hat{p}_i) = \Phi^{-1}(\hat{p}_i),$$

and is the value on the x-axis of a standard normal curve, at which the area to its left and under the curve is $\hat{p}_i$. Let $P_i(p_i) = \Phi^{-1}(p_i)$, then $E[P_i(\hat{p}_i)] = P_i(p_i)$, $\operatorname{var}[P_i(\hat{p}_i)] = [1/\phi\{\Phi^{-1}(p_i)\}]^2 p_i(1-p_i)/n$ and $\operatorname{covar}[P_i(\hat{p}_i), P_j(\hat{p}_j)] = 1/\phi\{\Phi^{-1}(p_i)\}\phi\{\Phi^{-1}(p_j)\}(-p_i p_j/n)$, $i, j = 0,1, \ldots, k$, $i \neq j$.
One can stack up the estimated probit transforms in a vector form and easily write the dispersion matrix.

2.11 Testing Linear Hypotheses

Consider the logit transformation as discussed in Section 2.10. We assume that

$$E(\hat{L}_0) = X\beta,$$

where **X** is a $k \times m$ matrix of rank r ($\leq m$), and β is a vector of m unknown parameters. We want to test the null hypothesis $H_0 : C\beta = \mathbf{d}$, where C is of order $p \times m$ and rank p ($\leq r$), and the parametric functions $C\beta$ are estimable. Let S be the estimated dispersion matrix:

$$S = \frac{1}{n}\left\{D\left(\frac{1}{\hat{p}_1}, \frac{1}{\hat{p}_2}, \ldots \frac{1}{\hat{p}_k}\right) + \frac{1}{\hat{p}_0} J_k\right\},$$

where $\mathbf{J}_k$ is a $k \times k$ matrix of all ones, and $D(1/\hat{\mathbf{p}}_1, 1/\hat{\mathbf{p}}_2, \ldots 1/\hat{\mathbf{p}}_k)$ is a diagonal matrix.

The lack of fit of the model is tested using the test statistic

$$T_1 = \hat{L}_0' S^{-1} \hat{L}_0 - \hat{\beta}' X' S^{-1} \hat{L}_C,$$

which has a χ^2 distribution with $k - r$ degrees of freedom, where $\hat{\beta}$ is any solution of the weighted normal equations:

$$X' S^{-1} X \hat{\beta} = X' S^{-1} \hat{L}_0.$$

It can be verified that $S^{-1} = n\{D(\hat{p}_1, \hat{p}_2, \ldots, \hat{p}_k) - \hat{p}\hat{p}'\}$, where $\hat{p}' = (\hat{p}_1, \hat{p}_2, \ldots, \hat{p}_k)$.

The null hypothesis $H_0 : C\beta = \mathbf{d}$ against the alternative hypothesis $H_A : C\beta \neq \mathbf{d}$ can be tested using the test statistic

$$T_2 = (C\hat{\beta} - \mathbf{d})' \{C(X' S^{-1} X)^- C'\}^{-1} (C\hat{\beta} - \mathbf{d}),$$

which has a χ^2 distribution with p degrees of freedom.

3

Generic Designs

3.1 Introduction

As discussed in Chapter 2, a linear model for the decomposition of the utility of a profile may include main effects, two-way, and higher-order interactions. The full factorial design may be modeled as

$$y_{x_1x_2\ldots x_m} = \mu + \sum_{i=1}^{m} \alpha_{x_i}^{A_i} + \sum_{\substack{i,j=1 \\ i\neq j}}^{m} \alpha_{x_ix_j}^{A_iA_j} + \sum_{\substack{i,j,k=1 \\ i\neq j\neq k\neq i}}^{m} \alpha_{x_ix_jx_k}^{A_iA_jA_k} + \ldots + \alpha_{x_1x_2\ldots x_m}^{A_1A_2\ldots A_m} + e_{x_1x_2\ldots x_m}$$

where $y_{x_1x_2\ldots x_m}$ is the response or transformed response of the profile $(x_1, x_2, \ldots, x_m)$; μ is the general mean; α_{xi}^{Ai} is the effect of factor A_i at x_i level; $\alpha_{xi\,xj}^{AiAj}$ is the effect of factors A_i and A_j at x_i and x_j levels, respectively; … ; and $e_{x_1x_2\ldots x_m}$ is the random error. In practice, subsets of terms are modeled in DCE (discrete choice experimentation) and CA (conjoint analysis). This chapter illustrates four types of models widely used in traditional DCE/CA. The four categories of models most widely assumed are models that include only brand effects, only attribute main effects, brand and attribute main effects, and brand and attribute main effects with selected two-way interactions. Illustrations of generic questions and experimental plans are provided. The last section that deals with brand and attribute main effects with selected two-way interactions illustrates estimation and hypothesis testing using simulated data.

While traditional CA data typically are analyzed using ordinary least squares (OLS), the Appendix provides an illustration of how random utility theory may also be applied to CA rating scale data using the categorical judgment model originally developed by Thurstone (1927).

3.2 Four Linear Models Used in CA and DCE

3.2.1 Brands-Only Models

Brands-only designs are ones by which respondents express preferences for concepts with no effort made to determine the relative contribution of the concept components to the overall evaluation of the concept. The descriptions may consist of brand names (such as brand names of automobiles), or they may be more complex, such as prospective promotional appeals or descriptions of prospective new products screened in a new product development process. The objective is determining the most preferred brand or concept. Partworths for concepts are modeled as

$$y_i = \mu + \alpha_i^B + e_i,$$

where α_i^B is the fixed partworth or utility of concept i on the nominal attribute B at level i, and e_i is the random error. Estimation also may be without intercept

$$y_i = b_i + e_i, \tag{3.1}$$

where $b_i = \mu + \alpha_i^B$.

3.2.2 Attributes-Only Models

With attributes-only designs, the utilities of concept profiles are decomposed into additive components:

$$y_{x_1 x_2 \ldots x_m} = \mu + \sum_{i=1}^{m} \alpha_{x_i}^{A_i} + e_{x_1 x_2 \ldots x_m},$$

where α_{xi}^{Ai} is the effect of factor A_i at x_i level, and $e_{x_1 x_2 \ldots x_m}$ is the random error. Attribute-only designs offer the advantage of enabling the researcher to determine the effect of changing levels of the respective attributes on the probability of choosing a concept profile. Their disadvantage is that they inevitably misspecify a choice situation in the sense that not all attributes relevant to respondents' choices may be included in a design specification.

3.2.3 Brand-Plus-Attributes Designs

Brand-plus-attribute designs combine brand-only and attribute-only specifications into a design in which brands are described in terms of attributes and a brand-specific parameter. The response to profiles is modeled as

$$y_{j x_1 x_2 \ldots x_m} = \mu + \alpha_j^B + \sum_{i=1}^{m} \alpha_{x_i}^{A_i} + e_{x_1 x_2 \ldots x_m},$$

where the model terms are defined in subsections 3.2.1 and 3.2.2. The brand-specific parameter picks up some misspecification of the attribute-only design. The attribute-specific components allow the researcher to determine the effect of changing characteristics of a concept profile, such as its price. Estimation may be with or without intercept as described with the brands-only model.

3.2.4 Brand-Plus-Attributes and Selected Two-Way Interactions Models

The brand-plus-attributes and selected two-way interactions models allow the effects of some attributes to depend on brand or selected other attributes. Accordingly, the response to a profile is modeled as

$$y_{lx_1x_2\dots x_m} = \mu + \alpha_l^B + \sum_{i=1}^{m} \alpha_{x_i}^{A_i} + \sum_{\substack{i,j=1 \\ i\neq j}}^{m} \alpha_{x_i x_j}^{A_i A_j} + e_{x_1x_2\dots x_m},$$

where the terms are defined in subsections 3.2.1 and 3.2.2. Estimation may be with or without intercept as described for the brands-only model. Including two-way interactions in a design allows the impact of an attribute, such as price, to depend on the brand having the price. For example, the interpretation of charging $41,000 for an automobile likely will vary depending on whether the automobile is a Chevrolet Aveo or a Cadillac Escalade.

3.3 Brands-Only Designs

With CA, the dependent variable consists of stated preferences for the concepts. Typically, preference is expressed using a rating scale with each concept rated. For example, the Pennsylvania mushroom growers association was interested in determining which of nine concepts were the most attractive to consumers. The nine concepts are shown in Exhibit 3A.1. Respondents rated each of the nine concepts using three scales: believability, uniqueness, and likelihood of purchase. In practice, these ratings would be analyzed using OLS at the individual level and the individual analyses aggregated to get aggregate partworth estimates. The Appendix to this chapter illustrates how ratings data may be analyzed using a random utility model (RUM)-based approach based on Thurstone's (1927) categorical judgment model. The illustrated approach provides a way to aggregate analysis that is consistent with the logit-based approaches we use to analyze DCE data.

With DCE, concepts are organized into blocks (choice sets), and respondents choose the most preferred concept in each choice set. For example, suppose an automobile manufacturer was interested in the impact of competitors coming into or leaving a market. Assume the relevant competitors were Ford Taurus, Chevrolet Malibu, Mitsubishi Lancer, Volvo C30, and Honda Accord. As shown in Table 3.1, an alternatives-only DCE for this scenario may be based on a Hadamard matrix $\mathbf{H}_8$ by dropping any three columns of H_8. (See Section 2.6 of Chapter 2 for an overview of Hadamard matrices.) Panel A of Table 3.1 presents $\mathbf{H}_8$. Panel B shows the results of dropping the first three columns of Panel A. Panel C shows the design plan for the experiment in which 1s indicate the brand is present in the choice set and 0s indicate that it is absent. All five brands are present in the first of the choice sets, brands E and G are present in the second choice set, and so forth. Note that in Panel C a no-choice (NC) alternative is added to each choice set.

Exhibit 3.1 illustrates how the first two choice sets for this illustrative DCE might be presented to respondents. In this example with only eight choice sets, respondents could be presented with all eight choice sets. However, if independence of choices across choice sets was desired, one choice set might be presented to each of eight groups in a between-subjects design. In other cases for which the DCE is part of a larger survey, it may not be possible to have all choice sets included in what is presented to any one respondent. In this case, a subset of choice sets may be included in the larger set of questions to which any single person responds. Analysis will be based on the aggregate of choices across individuals.

Table 3.2 shows hypothetical choice proportions for the design shown in Table 3.1, Panel C. In this hypothetical example, when given the choice of all five brands and no choice, 12% of respondents choose Ford Taurus and Chevrolet Malibu, 21% choose Mitsubishi Lancer, 29% choose Volvo C30, 26% choose Honda Accord, and 1% choose the no-choice option. When given the choice between Chevrolet Malibu, Volvo C30, and no choice, 28% choose Chevrolet Malibu, 70% choose Volvo C30, and 2% choose the no-choice option. These proportions sum to 100% (subject to rounding error). Note that in the present example we assume that Equation 3.1 correctly specifies consumers' composition rules, and the brand partworths b are measured without error. The "true" choice proportions for concept i in choice set S are given by

$$p_{j\in S} = e^{b_i} / \sum_{j\in S} e^{b_j}, \tag{3.2}$$

where b_i is the partworth for concept i. Note that the choice proportions given by Equation 3.2 will be nonzero.

Table 3.3 presents the dependent variable and design matrix for the alternatives-only design plan shown in Table 3.1, Panel C. The dependent variable is the logit of the proportion choosing each alternative in the set to the proportion choosing the no-choice option in Table 3.2. Thus, there is one observation for each concept in a choice set. Each nonzero element of the design plan contributes one row to the design matrix and the dependent

TABLE 3.1
Hadamard Design for Alternatives-Only DCE

A								B						C						
Hadamard H_8								Last Five Columns H_8					Choice	Design Plan						Set
A	B	C	D	E	F	G	H	D	E	F	G	H	Set	D	E	F	G	H	NC	Size
1	1	1	1	1	1	1	1	1	1	1	1	1	1	1	1	1	1	1	1	6
1	–1	1	–1	1	–1	1	–1	–1	1	–1	1	–1	2	0	1	0	1	0	1	3
1	1	–1	–1	1	1	–1	–1	–1	1	1	–1	–1	3	0	1	1	0	0	1	3
1	–1	–1	1	1	–1	–1	1	1	1	–1	–1	1	4	1	1	0	0	1	1	4
1	1	1	1	–1	–1	–1	–1	1	–1	–1	–1	–1	5	1	0	0	0	0	1	2
1	–1	1	–1	–1	1	–1	1	–1	–1	1	–1	1	6	0	0	1	0	1	1	3
1	1	–1	–1	–1	–1	1	1	–1	–1	–1	1	1	7	0	0	0	1	1	1	3
1	–1	–1	1	–1	1	1	–1	1	–1	1	1	–1	8	1	0	1	1	0	1	4

EXHIBIT 3.1

Illustrative Presentation of First Two Choice Sets from Table 3.1

Suppose you were deciding to purchase a new automobile. If the only available alternatives were the ones shown in each of the choice sets that follow, which automobile would you be most likely to purchase? Indicate the one you would be most likely to purchase by checking the box below the brand name. If you would purchase none of the available alternatives, check the box under "If these were the only alternatives, I would not purchase or would delay purchase."

Choice Set 1

Ford Taurus	Chevrolet Malibu	Mitsubishi Lancer	Volvo C30	Honda Accord
☐	☐	☐	☐	☐

If these were the only available options, I would not purchase or would delay purchase.

☐

Choice Set 2

Chevrolet Malibu	Volvo C30
☐	☐

If these were the only available options, I would not purchase or would delay purchase.

☐

TABLE 3.2

Choice Proportions for Hypothetical Alternatives-Only DCE

	Brand					
Choice Set	Ford Taurus	Chevrolet Malibu	Mitsubishi Lancer	Volvo C30	Honda Accord	NC
1	0.12	0.12	0.21	0.29	0.26	0.01
2	0	0.28	0	0.7	0	0.02
3	0	0.35	0.63	0	0	0.03
4	0.23	0.23	0	0	0.52	0.02
5	0.93	0	0	0	0	0.07
6	0	0	0.44	0	0.54	0.02
7	0	0	0	0.52	0.47	0.02
8	0.19	0	0.34	0.46	0	0.01

TABLE 3.3

Dependent Variable $-\ln(p_{ij}/p_{i,nc})$ and Design Matrix for Design Plan, Table 3.1, Panel C

	Y	*X*				
Choice Set	**ln $(p_{ij}/p_{i,nc})$**	**Ford Taurus**	**Chevrolet Malibu**	**Mitsubishi Lancer**	**Volvo C30**	**Honda Accord**
1	2.6	1	0	0	0	0
1	2.6	0	1	0	0	0
1	3.2	0	0	1	0	0
1	3.5	0	0	0	1	0
1	3.4	0	0	0	0	1
2	2.6	0	1	0	0	0
2	3.5	0	0	0	1	0
3	2.6	0	1	0	0	0
3	3.2	0	0	1	0	0
4	2.6	1	0	0	0	0
4	2.6	0	1	0	0	0
4	3.4	0	0	0	0	1
5	2.6	1	0	0	0	0
6	3.2	0	0	1	0	0
6	3.4	0	0	0	0	1
7	3.5	0	0	0	1	0
7	3.4	0	0	0	0	1
8	2.6	1	0	0	0	0
8	3.2	0	0	1	0	0
8	3.5	0	0	0	1	0

variable. For example, choice set 1 contributes 5 rows, choice set 2 contributes 2 rows, choice set 4 contributes 3 rows, and so forth for a total of 20 rows (or observations). There are five parameters (the utilities) to be estimated, leaving 15 degrees of freedom. The utilities of the concepts are estimated by regressing the variables of the columns of the design matrix **X** on the dependent variable. Note that this regression equation is without intercept. The values for the partworths (regression coefficients) are determined by regressing the design matrix **X** on **Y** (Table 3.3). In this example, the partworths are 2.6, 2.6, 3.2, 3.5, and 3.4 for Ford Taurus, Chevrolet Malibu, Mitsubishi Lancer, Volvo C30, and Honda Accord, respectively.

We now consider the consequences of Volvo C30 entering a market consisting of Ford Taurus, Chevrolet Malibu, Mitsubishi Lancer, and Honda Accord. Table 3.4 shows the calculations using Equation 3.2 to arrive at estimated market shares for markets consisting of the four brands (Ford Taurus, Chevrolet Malibu, Mitsubishi Lancer, and Honda Accord) and a market consisting of these four brands with the addition of Volvo C30. The estimated market shares for the four- and five-brand markets are given in the rows

TABLE 3.4

Market Share Estimates for Four- and Five-Brand Markets

	Ford Taurus	Chevrolet Malibu	Mitsubishi Lancer	Honda Accord	Volvo C30	Sum
Four-Brand Market						
Utilities *b*	2.6	2.6	3.2	3.4		
exp(b_i)	13.46	13.46	24.53	29.96		81.42
p	.17	.17	.30	.37		1.00
Five-Brand Market						
Utilities *b*	2.6	2.6	3.2	3.4	3.5	
exp(b_i)	13.46	13.46	24.53	29.96	33.12	114.54
p	.12	.12	.21	.26	.29	1.00
Change in share	−0.05	−0.05	−0.09	−0.11	0.29	0.00

labeled *p*, and these are given by $p_{j\in S} = e^{b_i} / \sum_{j\in S} e^{b_j}$. Adding Volvo C30 to the four-brand market results in Volvo C30 picking up 29% market share. This share is taken from the four initial brands in the market, with 5% coming from Ford Taurus and Chevrolet Malibu, 9% taken from Mitsubishi Lancer, and 11% taken from Honda Accord. In accordance with the IIA property that characterizes the multinomial logit (MNL) model, these shares are taken in proportion to the brand shares in the four-brand market, that is, 5% = .29 × .17; 9% = .29 × .30, and 11% = .29 × .37 (subject to rounding error).

3.4 Attributes-Only Designs

Attributes-only designs present concept profiles consisting of a specific combination of indicated levels on *m* specified attributes $m \geq 2$. Among the situations for which attribute-only designs are used are ones in which all attribute combinations refer to the same brand, to screen attributes for "importance," or in new product development situations when the objective is to determine which combinations of attributes are most preferred by prospective customers.

For example, price is an attribute of a product concept, such as an automobile. Other attributes might include color (white, red, blue, green); seat covering (vinyl, leather); number of doors (two, four); size of engine (four, six, eight cylinders); type of transmission (stick, automatic); and feature packages (say four combinations of tires, soundproofing, sound system, navigation).

Suppose of the many attributes that might describe automobiles there were four attributes that the firm wanted to evaluate: five levels of price, four of transmission, three of engine type, and two of number of doors. There are 120 combinations of these four attributes, each describing a possible product concept in a full factorial experimental design. By assuming that interactions are negligible, however, it is possible to collect information necessary for estimating all main-effect partworths using 25 profiles. Table 3.5 shows 25 profiles for one design in 25 runs that was generated using the ORTHOPLAN module available in SPSS (2008). This is a fractional factorial plan of an asymmetric factorial experiment and is more general than the one mentioned in Chapter 2. The alias structure of this plan is more complicated than the fractional plans of symmetrical factorial experiments discussed. In a CA design, each of these 25 profiles would be presented to respondents, who would express their preference using a rating scale. For example, the rating scale might ask the respondent to indicate likelihood of purchase.

In a DCE, the 25 profiles may be organized into five sets of five profiles. Assuming a base option was included with each of the five choice sets, each of the five choice sets would contribute five logits to an analysis for a total of 25 observations. There are (5–1) + (4–1) + (3–1) + (2–1) + 1 = 11 parameters to estimate, including a mean. Exhibit 3.2 shows the first choice set of Table 3.5 in which five attributes-only profiles of automobiles are described in terms of four attributes: price, transmission type, engine, and number of doors.

Table 3.6 presents the design matrix **X** using "effects" coding (1, 0, –1) for the design shown in Table 3.5, Panel A. With effects coding, the attribute effects are constrained to sum to 0.00, that is, $\sum_{j=1}^{S_i} \alpha_j^{A_i} = 0$, where attribute A_i has s_i levels. That is, the effect of any one level of an attribute is equal to minus the sum of the remaining attributes. For example, in Table 3.6 the $41,000 price is not included as a column. For the third profile, we code the $41,000 price as –1 for the other four price effects. Number of doors is coded as: 2 doors = 1, 4 doors = –1

3.5 Brands-Plus-Attributes Designs

Brands-plus-attributes designs allow estimation of attribute effects and a brand-specific parameter. The attribute effects capture the effects of changes of attribute levels on the probability of choosing a profile. The brand-specific attribute captures partworths attributable to the brand name. In addition, brand-specific attributes capture some effects of misspecification of the attributes. For example, if some attribute or attribute level associated with a brand were left out of the design, the estimate of the brand-specific coefficient would pick up some systematic variation due to misspecification of the model.

Table 3.7 shows a design for five brands and the same attributes used in the Section 3.4. This design also has 25 runs, or profiles, and also was generated by the ORTHOPLAN SPSS (2008) procedure. Brands are coded as (1) Ford

TABLE 3.5

Attributes-Only Plan in 25 Runs for Four Attributes with Two to Five Levels

A					B				
Profile	Price	Transmission	Engine	Doors	Choice Set	Price	Transmission	Engine	Doors
1	3	1	1	2	1	$31,000	Manual 5 speed	4 cylinder	4 door
2	3	1	2	1	1	$31,000	Manual 5 speed	6 cylinder	2 door
3	5	2	3	2	1	$41,000	Automatic 5 speed	8 cylinder	4 door
4	2	1	1	1	1	$26,000	Manual 5 speed	4 cylinder	2 door
5	3	3	3	1	1	$31,000	Manual 6 speed	8 cylinder	2 door
6	2	1	3	2	2	$26,000	Manual 5 speed	8 cylinder	4 door
7	3	4	2	1	2	$31,000	Automatic 6 speed	6 cylinder	2 door
8	4	4	1	1	2	$36,000	Automatic 6 speed	4 cylinder	2 door
9	2	4	2	2	2	$26,000	Automatic 6 speed	6 cylinder	4 door
10	5	1	2	1	2	$41,000	Manual 5 speed	6 cylinder	2 door
11	2	3	1	1	3	$26,000	Manual 6 speed	4 cylinder	2 door
12	1	4	3	1	3	$21,000	Automatic 6 speed	8 cylinder	2 door

13	1	3	1	2	3	$21,000	Manual 6 speed	4 cylinder	4 door
14	2	2	2	1	3	$26,000	Automatic 5 speed	6 cylinder	2 door
15	5	1	1	1	3	$41,000	Manual 5 speed	4 cylinder	2 door
16	1	1	2	2	4	$21,000	Manual 5 speed	6 cylinder	4 door
17	1	2	2	1	4	$21,000	Automatic 5 speed	6 cylinder	2 door
18	5	4	1	2	4	$41,000	Automatic 6 speed	4 cylinder	4 door
19	5	3	2	1	4	$41,000	Manual 6 speed	6 cylinder	2 door
20	3	2	1	2	4	$31,000	Automatic 5 speed	4 cylinder	4 door
21	4	2	1	1	5	$36,000	Automatic 5 speed	4 cylinder	2 door
22	1	1	1	1	5	$21,000	Manual 5 speed	4 cylinder	2 door
23	4	1	2	2	5	$36,000	Manual 5 speed	6 cylinder	4 door
24	4	3	2	2	5	$36,000	Manual 6 speed	6 cylinder	4 door
25	4	1	3	1	5	$36,000	Manual 5 speed	8 cylinder	2 door

EXHIBIT 3.2
Illustration of One Choice Set of an Attributes-Only Design

The following are five automobile profiles (G, H, I, J, and K) described in terms of price, type of transmission, type of engine, and number of doors. Indicate (by placing a mark in the box below the profile) the profile you would be most likely to choose if these were the only options available. If you would not purchase an automobile (or would delay your purchase) if these were the only options available, check the box provided on the bottom right.

	Automobile				
	G	H	I	J	K
Price	$31,000	$31,000	$41,000	$26,000	$31,000
Transmission	Manual 5 speed	Manual 5 speed	Automatic 5 speed	Manual 5 speed	Manual 6 speed
Engine	4 cylinder	6 cylinder	8 cylinder	4 cylinder	8 cylinder
Doors	4 door	2 door	4 door	2 door	2 door
	☐	☐	☐	☐	☐
If these were the only available choices, I would choose none of them.					☐

Taurus, (2) Chevrolet Malibu, (3) Mitsubishi Lancer, (4) Volvo C30, and (5) Honda Accord. As in the previous attributes-only example, the respective attributes and levels are price (5), transmission type (4), engine type (3), and number of doors (2). In a CA study, each of the 25 profiles would be rated on a relevant attribute, such as likelihood of purchase.

Table 3.8 provides coding for the design in Table 3.7. Note that the coding in Table 3.8 includes all brands, so estimation will be without an overall mean. A binary coding scheme is used for the remaining attributes. Binary coding of a nominal/categorical variable involves choosing an excluded reference category for each nominal/categorical variable. With this coding, the effect of the profile described by the excluded category of each attribute is given by the brand effect, which is equal to the mean rating for the brand. The parameter estimates of the remaining categories are the differential effects of the remaining categories compared to the brand effect. For example, the parameters for price levels $21,000 through $36,000 would be expected to be positive as they are compared to the effect of the $41,000 price level for each brand.

3.6 Brands, Attributes, and Interaction Design

Designs with interactions allow the partworth of attribute levels to depend on the level of other attributes. For example, consumers might classify automobiles into "prestige" versus "economy" categories. There may be

TABLE 3.6
Attributes-Only Design Matrix (Contrast Coding)

	Price				Transmission Type			Engine Type		Doors
Mean	$21,000	$26,000	$31,000	$36,000	Manual 5	Automatic 5	Manual 6	4 Cyl.	6 Cyl.	2 Doors
1	0	0	1	0	1	0	0	1	0	–1
1	0	0	1	0	1	0	0	0	1	1
1	–1	–1	–1	–1	0	1	0	–1	–1	–1
1	0	1	0	0	1	0	0	1	0	1
1	0	0	1	0	0	0	1	–1	–1	1
1	0	1	0	0	1	0	0	–1	–1	–1
1	0	0	1	0	–1	–1	–1	0	1	1
1	0	0	0	1	–1	–1	–1	1	0	1
1	0	1	0	0	–1	–1	–1	0	1	–1
1	–1	–1	–1	–1	1	0	0	0	1	1
1	0	1	0	0	0	0	1	1	0	1
1	1	0	0	0	–1	–1	–1	–1	–1	1
1	1	0	0	0	0	0	1	1	0	–1
1	0	1	0	0	0	1	0	0	1	1
1	–1	–1	–1	–1	1	0	0	1	0	1
1	1	0	0	0	1	0	0	0	1	–1
1	1	0	0	0	0	1	0	0	1	1
1	–1	–1	–1	–1	–1	–1	–1	1	0	–1
1	–1	–1	–1	–1	0	0	1	0	1	1
1	0	0	1	0	0	1	0	1	0	–1

(Continued)

TABLE 3.6 (Continued)

Attributes-Only Design Matrix (Contrast Coding)

	Price				Transmission Type			Engine Type		Doors
Mean	$21,000	$26,000	$31,000	$36,000	Manual 5	Automatic 5	Manual 6	4 Cyl.	6 Cyl.	2 Doors
1	0	0	0	1	0	1	0	1	0	1
1	1	0	0	0	1	0	0	1	0	1
1	0	0	0	1	1	0	0	0	1	–1
1	0	0	0	1	0	0	1	0	1	–1
1	0	0	0	1	1	0	0	–1	–1	1

TABLE 3.7

Design for Five Brands with Four Attributes

Choice set	**1**	**1**	**1**	**1**	**1**	**2**	**2**	**2**	**2**	**2**	**3**	**3**	**3**	**3**	**3**	**4**	**4**	**4**	**4**	**4**	**5**	**5**	**5**	**5**	**5**
Brand	5	1	4	1	1	3	2	4	3	1	4	1	3	2	2	3	5	2	4	5	2	3	5	4	5
Price	1	5	2	3	2	4	4	5	1	4	3	1	3	5	2	2	3	1	1	4	3	5	2	4	5
Transmission	3	2	4	4	1	4	1	1	2	3	3	1	1	1	3	1	1	4	1	1	2	3	2	2	4
Engine	3	3	3	2	1	1	3	2	2	2	1	1	3	1	2	2	1	1	2	2	2	1	1	1	2
Doors	2	1	1	2	2	2	2	2	1	1	1	1	1	1	1	1	1	1	2	1	2	2	2	1	1

TABLE 3.8

Coding of Design Matrix

	Design Matrix						Brand				Price				Transmission			Engine		Doors
Profile	Brand	Price	Transmission	Engine	Doors	Logit	Chevrolet Malibu	Ford Taurus	Mitsubishi Lancer	Volvo C30	$21,000	$26,000	$31,000	$36,000	Manual 5 Speed	Manual 6 Speed	Automatic 5 Speed	4 Cylinder	6 Cylinder	2 Doors
1	5	1	3	3	2	$L_{11/\text{base}}$	−1	−1	−1	−1	1	0	0	0	0	0	1	−1	−1	−1
2	1	5	2	3	1	$L_{21/\text{base}}$	1	0	0	0	−1	−1	−1	−1	0	1	0	−1	−1	1
3	4	2	4	3	1	$L_{31/\text{base}}$	0	0	0	1	0	1	0	0	−1	−1	−1	−1	−1	1
4	1	3	4	2	2	$L_{41/\text{base}}$	1	0	0	0	0	0	1	0	−1	−1	−1	0	1	−1
5	1	2	1	1	2	$L_{51/\text{base}}$	1	0	0	0	0	1	0	0	1	0	0	1	0	−1
6	3	4	4	1	2	$L_{12/\text{base}}$	0	0	1	0	0	0	0	1	−1	−1	−1	1	0	−1
7	2	4	1	3	2	$L_{22/\text{base}}$	0	1	0	0	0	0	0	1	1	0	0	−1	−1	−1
8	4	5	1	2	2	$L_{32/\text{base}}$	0	0	0	1	−1	−1	−1	−1	1	0	0	0	1	−1
9	3	1	2	2	1	$L_{42/\text{base}}$	0	0	1	0	1	0	0	0	0	1	0	0	1	1
10	1	4	3	2	1	$L_{52/\text{base}}$	1	0	0	0	0	0	0	1	0	0	1	0	1	1
11	4	3	3	1	1	$L_{13/\text{base}}$	0	0	0	1	0	0	1	0	0	0	1	1	0	1
12	1	1	1	1	1	$L_{23/\text{base}}$	1	0	0	0	1	0	0	0	1	0	0	1	0	1
13	3	3	1	3	1	$L_{33/\text{base}}$	0	0	1	0	0	0	1	0	1	0	0	−1	−1	1
14	2	5	1	1	1	$L_{43/\text{base}}$	0	1	0	0	−1	−1	−1	−1	1	0	0	1	0	1
15	2	2	3	2	1	$L_{53/\text{base}}$	0	1	0	0	0	1	0	0	0	0	1	0	1	1
16	3	2	1	2	1	$L_{14/\text{base}}$	0	0	1	0	0	1	0	0	1	0	0	0	1	1
17	5	3	1	1	1	$L_{24/\text{base}}$	−1	−1	−1	−1	0	0	1	0	1	0	0	1	0	1
18	2	1	4	1	1	$L_{34/\text{base}}$	0	1	0	0	1	0	0	0	−1	−1	−1	1	0	1

19	4	1	1	2	2	L_{44}/base	0	0	0	1	1	0	0	0	1	0	0	0	1	−1
20	5	4	1	2	1	L_{54}/base	−1	−1	−1	−1	0	0	0	1	1	0	0	0	1	1
21	2	3	2	2	2	L_{15}/base	0	1	0	0	0	0	1	0	0	1	0	0	1	−1
22	3	5	3	1	2	L_{25}/base	0	0	1	0	−1	−1	−1	−1	0	0	1	1	0	−1
23	5	2	2	1	2	L_{35}/base	−1	−1	−1	−1	0	1	0	0	0	1	0	1	0	−1
24	4	4	2	1	1	L_{45}/base	0	0	0	1	0	0	0	1	0	1	0	1	0	1
25	5	5	4	2	1	L_{55}/base	−1	−1	−1	−1	−1	−1	−1	−1	−1	−1	−1	0	1	1

expectations regarding the features that are typical of offerings in the luxury versus economy categories. Offering a feature that exceeds expectations for a category may result in larger choice probability than would be the case in the absence of such expectations. Conversely, failing to meet expectations may result in smaller choice probabilities than would be the case in the absence of such expectations. For example, offering power doors, windows, and mirrors may be perceived as a substantial incremental benefit in an economy car but as no incremental benefit in a luxury car. Conversely, their absence may be perceived as a substantial loss in a luxury car but not in an economy car. In this case, we would expect to find interactions of product category by attribute.

It also may be the case that there are groups in the populations that differ in their partworths for attribute levels. For example, there might be economy and prestige segments in a market that differentially value levels of product offerings (e.g., price levels). People in the economy segment may be price sensitive in the price range characterizing automobiles in the economy category but be insensitive to price levels in the price range characterizing prestige category automobiles because they are equally unlikely to purchase an auto from this category at any price characterizing the category. Members of the prestige segment, on the other hand, may be price sensitive in the range characterizing offerings in the prestige category but price insensitive to price changes in the range of price levels characterizing the economy category because they are unlikely to buy brands from the economy category.

As discussed in Chapter 2, a full factorial design allows all interactions between attributes to be estimated. When there are many attributes or levels, full factorial designs result in many profiles, more than may be presented to a single respondent. Even when respondents are presented subsets of alternatives, collecting observations for all profiles requires larger samples than are feasible.

However, it usually is unnecessary to estimate interactions higher than three way. In DCE contexts, choices are the outcome of human information processing, and it may be assumed that the impacts of such higher-order interactions on choice are negligible. Even if higher-order interactions are marginally present, it is difficult to give meaningful interpretation to them. On the other hand, the presence of some significant two-way and some three-way interactions is plausible. For example, as noted, it is plausible that there will be interactions between brand or product category and other attributes or between price level and other product attributes. Regarding three-way interactions, it often is plausible that two-way interactions will differ across groups in a population, such as market segments. For example, in the following illustrative example, it is assumed there are two-way interactions between product category and price-level partworths. Furthermore, it is assumed that these two-way interactions differ in the economy and prestige market segments. The economy segment is price sensitive in the price range characterizing economy cars and price insensitive in the price range characterizing prestige cars. The reverse is the case for the prestige segment. This is a case of an interaction of segment by product category by price level.

Table 3.9 provides a 5 x 2^3 experimental design for product profiles of hypothetical economy and prestige category automobiles. The price levels of the two categories differ, with the levels of the economy category ranging from \$21,000 to \$41,000 and the levels of the prestige category ranging from \$51,000 to \$91,000. The remaining attributes are engine type (six vs. eight cylinder) and transmission type (manual vs. automatic and five speed vs. six speed). Panel A of the design provides the experimental design. Panel B provides the values corresponding to the design levels. Panel C provides contrast coding for the attribute levels. Note that price levels are "nested" within product category, which is necessary because the values for price differ in each category. That is, the price effects are estimated as main effects on the mean effect of category.

The design is constructed by using the methods discussed in Chapter 2. Two one-half fractions of four profiles are formed with the three attributes Eng (engine), M/A (manual/automatic), and 5/6 (five speed/six speed). One-half fraction is taken at each of the five price levels of economy cars, and the other fraction is taken at each of the price levels of luxury cars. The result is a design in which a complete design for two-level attributes is present at each price level in category (and each customer segment). It also is possible to estimate two-way interactions between two-level attributes. The 40 runs are organized into eight choice sets of size five.

Note in Table 3.9 that an interaction is included between the two attributes of transmission. This approach provides an alternative coding to the "superfactor" coding for transmission illustrated in the previous section. The two approaches are equivalent from a statistical standpoint, for example, the same number of parameters (three) is estimated, and the same proportion of variance will be accounted for with each approach. The interaction coding approach has the advantage that a test for the significance of the two-way interaction is straightforward. If the test proves insignificant, usually the interaction term may be eliminated and the more parsimonious main-effects-only modeling of transmission effects adopted. On the other hand, the superfactor approach is most convenient when some combinations of the factors are absent. For example, if a six-speed version of the manual transmission were not offered, then the three combinations of transmission offered should be modeled as a superfactor.

Tables 3.10a and 3.10b provide the coding for a design having two market segments. The coding for the profiles is carried over from Table 3.9 with the addition of an overall mean. The design is repeated for each segment. Tables 3.10a and 3.10b also provide illustrative data for the two population segments in the example.

TABLE 3.9

Brand, Attributes, with Interactions Design

	(A) Design Levels					(B) Design Values					(C) Coding												
												Economy Price				Prestige Price							
Ch. Set	Cat	Pr	Eng	M/A	5/6	Cat	Pr	Eng	M/A	5/6	Cat	P1	P2	P3	P4	P1	P2	P3	P4	Eng	M/A	5/6	Int
1	1	1	1	1	2	E	$21k	6 cyl.	Man.	6 sp.	–1	1	0	0	0	0	0	0	0	–1	–1	1	–1
1	1	1	1	2	1	E	$21k	6 cyl.	Auto.	5 sp.	–1	1	0	0	0	0	0	0	0	–1	1	–1	–1
1	1	1	2	1	1	E	$21k	8 cyl.	Man.	5 sp.	–1	1	0	0	0	0	0	0	0	1	–1	–1	1
1	1	1	2	2	2	E	$21k	8 cyl.	Auto.	6 sp.	–1	1	0	0	0	0	0	0	0	1	1	1	1
1	1	2	1	1	2	E	$26k	6 cyl.	Man.	6 sp.	–1	0	1	0	0	0	0	0	0	–1	–1	1	–1
2	1	2	1	2	1	E	$26k	6 cyl.	Auto.	5 sp.	–1	0	1	0	0	0	0	0	0	–1	1	–1	–1
2	1	2	2	1	1	E	$26k	8 cyl.	Man.	5 sp.	–1	0	1	0	0	0	0	0	0	1	–1	–1	1
2	1	2	2	2	2	E	$26k	8 cyl.	Auto.	6 sp.	–1	0	1	0	0	0	0	0	0	1	1	1	1
2	1	3	1	1	2	E	$31k	6 cyl.	Man.	6 sp.	–1	0	0	1	0	0	0	0	0	–1	–1	1	–1
2	1	3	1	2	1	E	$31k	6 cyl.	Auto.	5 sp.	–1	0	0	1	0	0	0	0	0	–1	1	–1	–1
3	1	3	2	1	1	E	$31k	8 cyl.	Man.	5 sp.	–1	0	0	1	0	0	0	0	0	1	–1	–1	1
3	1	3	2	2	2	E	$31k	8 cyl.	Auto.	6 sp.	–1	0	0	1	0	0	0	0	0	1	1	1	1
3	1	4	1	1	2	E	$36k	6 cyl.	Man.	6 sp.	–1	0	0	0	1	0	0	0	0	–1	–1	1	–1
3	1	4	1	2	1	E	$36k	6 cyl.	Auto.	5 sp.	–1	0	0	0	1	0	0	0	0	–1	1	–1	–1
3	1	4	2	1	1	E	$36k	8 cyl.	Man.	5 sp.	–1	0	0	0	1	0	0	0	0	1	–1	–1	1
4	1	4	2	2	2	E	$36k	8 cyl.	Auto.	6 sp.	–1	0	0	0	1	0	0	0	0	1	1	1	1
4	1	5	1	1	2	E	$41k	6 cyl.	Man.	6 sp.	–1	–1	–1	–1	–1	0	0	0	0	–1	–1	1	–1
4	1	5	1	2	1	E	$41k	6 cyl.	Auto.	5 sp.	–1	–1	–1	–1	–1	0	0	0	0	–1	1	–1	–1
4	1	5	2	1	1	E	$41k	8 cyl.	Man.	5 sp.	–1	–1	–1	–1	–1	0	0	0	0	1	–1	–1	1
4	1	5	2	2	2	E	$41k	8 cyl.	Auto.	6 sp.	–1	–1	–1	–1	–1	0	0	0	0	1	1	1	1
5	2	1	1	1	1	P	$51k	6 cyl.	Man.	5 sp.	1	0	0	0	0	1	0	0	0	–1	–1	–1	1
5	2	1	1	2	2	P	$51k	6 cyl.	Auto.	6 sp.	1	0	0	0	0	1	0	0	0	–1	1	1	1

5	2	1	2	1	2	P	$51k	8 cyl.	Man.	6 sp.	1	0	0	0	0	1	0	0	0	1	–1	1	–1
5	2	1	2	2	1	P	$51k	8 cyl.	Auto.	5 sp.	1	0	0	0	0	1	0	0	0	1	1	–1	–1
5	2	2	1	1	1	P	$61k	6 cyl.	Man.	5 sp.	1	0	0	0	0	0	1	0	0	–1	–1	–1	1
6	2	2	1	2	2	P	$61k	6 cyl.	Auto.	6 sp.	1	0	0	0	0	0	1	0	0	–1	1	1	1
6	2	2	2	1	2	P	$61k	8 cyl.	Man.	6 sp.	1	0	0	0	0	0	1	0	0	1	–1	1	–1
6	2	2	2	2	1	P	$61k	8 cyl.	Auto.	5 sp.	1	0	0	0	0	0	1	0	0	1	1	–1	–1
6	2	3	1	1	1	P	$71k	6 cyl.	Man.	5 sp.	1	0	0	0	0	0	0	1	0	–1	–1	–1	1
6	2	3	1	2	2	P	$71k	6 cyl.	Auto.	6 sp.	1	0	0	0	0	0	0	1	0	–1	1	1	1
7	2	3	2	1	2	P	$71k	8 cyl.	Man.	6 sp.	1	0	0	0	0	0	0	1	0	1	–1	1	–1
7	2	3	2	2	1	P	$71k	8 cyl.	Auto.	5 sp.	1	0	0	0	0	0	0	1	0	1	1	–1	–1
7	2	4	1	1	1	P	$81k	6 cyl.	Man.	5 sp.	1	0	0	0	0	0	0	0	1	–1	–1	–1	1
7	2	4	1	2	2	P	$81k	6 cyl.	Auto.	6 sp.	1	0	0	0	0	0	0	0	1	–1	1	1	1
7	2	4	2	1	2	P	$81k	8 cyl.	Man.	6 sp.	1	0	0	0	0	0	0	0	1	1	–1	1	–1
8	2	4	2	2	1	P	$81k	8 cyl.	Auto.	5 sp.	1	0	0	0	0	0	0	0	1	1	1	–1	–1
8	2	5	1	1	1	P	$91k	6 cyl.	Man.	5 sp.	1	0	0	0	0	–1	–1	–1	–1	–1	–1	–1	1
8	2	5	1	2	2	P	$91k	6 cyl.	Auto.	6 sp.	1	0	0	0	0	–1	–1	–1	–1	–1	1	1	1
8	2	5	2	1	2	P	$91k	8 cyl.	Man.	6 sp.	1	0	0	0	0	–1	–1	–1	–1	1	–1	1	–1
8	2	5	2	2	1	P	$91k	8 cyl.	Auto.	5 sp.	1	0	0	0	0	–1	–1	–1	–1	1	1	–1	–1

Cat = category, 1 = Economy (E), 2 = Prestige (P); Pr = Price, 1 = $21,000 and 5 = $41,000 in the Economy category, and 1 = $51,000, 5 = $91,000 in the prestige category; Eng = Engine, 1 = 6 cylinder, 2 = 8 cylinder; M/A = Transmission type, M = Manual, A = Automatic; 5/6 = Gears, 5 = 5 levels, 6 = 6 levels; Int = Interaction between M/A and 5/6.

TABLE 3.10A

Upper Half of Design Matrix (Economy Segment)

		(A) True Partworths																						
		0.00	−0.25	0.25	0.75	0.25	−0.13	−0.38	0.01	0.01	0.00	−0.01	−0.01	−0.01	0.00	0.01	1.59	1.05	−0.39	−1.08	0.07	0.10	0.30	0.25
Choice			Cat.	Cat.	Economy Price				Prestige Price				Economy Price				Prestige Price							
Set	Logits (L)	Mean	(econ)	(pres)	P1	P2	P3	P4	P1	P2	P3	P4	P1	P2	P3	P4	P1	P2	P3	P4	Eng	M/A	5/6	Int
1	0.88	1	−1	0	1	0	0	0	0	0	0	0	0	0	0	0	0	0	0	0	−1	−1	1	−1
1	0.48	1	−1	0	1	0	0	0	0	0	0	0	0	0	0	0	0	0	0	0	−1	1	−1	−1
1	0.92	1	−1	0	1	0	0	0	0	0	0	0	0	0	0	0	0	0	0	0	1	−1	−1	1
1	1.72	1	−1	0	1	0	0	0	0	0	0	0	0	0	0	0	0	0	0	0	1	1	1	1
1	0.38	1	−1	0	0	1	0	0	0	0	0	0	0	0	0	0	0	0	0	0	−1	−1	1	−1
2	−0.02	1	−1	0	0	1	0	0	0	0	0	0	0	0	0	0	0	0	0	0	−1	1	−1	−1
2	0.42	1	−1	0	0	1	0	0	0	0	0	0	0	0	0	0	0	0	0	0	1	−1	−1	1
2	1.22	1	−1	0	0	1	0	0	0	0	0	0	0	0	0	0	0	0	0	0	1	1	1	1
2	0.00	1	−1	0	0	0	1	0	0	0	0	0	0	0	0	0	0	0	0	0	−1	−1	1	−1
2	−0.40	1	−1	0	0	0	1	0	0	0	0	0	0	0	0	0	0	0	0	0	−1	1	−1	−1
3	0.04	1	−1	0	0	0	1	0	0	0	0	0	0	0	0	0	0	0	0	0	1	−1	−1	1
3	0.84	1	−1	0	0	0	1	0	0	0	0	0	0	0	0	0	0	0	0	0	1	1	1	1
3	−0.25	1	−1	0	0	0	0	1	0	0	0	0	0	0	0	0	0	0	0	0	−1	−1	1	−1
3	−0.65	1	−1	0	0	0	0	1	0	0	0	0	0	0	0	0	0	0	0	0	−1	1	−1	−1
3	−0.21	1	−1	0	0	0	0	1	0	0	0	0	0	0	0	0	0	0	0	0	1	−1	−1	1
4	0.59	1	−1	0	0	0	0	1	0	0	0	0	0	0	0	0	0	0	0	0	1	1	1	1
4	−0.36	1	−1	0	−1	−1	−1	−1	0	0	0	0	0	0	0	0	0	0	0	0	−1	−1	1	−1
4	−0.76	1	−1	0	−1	−1	−1	−1	0	0	0	0	0	0	0	0	0	0	0	0	−1	1	−1	−1
4	−0.32	1	−1	0	−1	−1	−1	−1	0	0	0	0	0	0	0	0	0	0	0	0	1	−1	−1	1
4	0.48	1	−1	0	−1	−1	−1	−1	0	0	0	0	0	0	0	0	0	0	0	0	1	1	1	1

5	−0.46	1	1	0	0	0	0	0	1	0	0	0	0	0	0	0	0	0	0	0	−1	−1	−1	1
5	0.34	1	1	0	0	0	0	0	1	0	0	0	0	0	0	0	0	0	0	0	−1	1	1	1
5	−0.22	1	1	0	0	0	0	0	1	0	0	0	0	0	0	0	0	0	0	0	1	−1	1	−1
5	−0.62	1	1	0	0	0	0	0	1	0	0	0	0	0	0	0	0	0	0	0	1	1	−1	−1
5	−0.46	1	1	0	0	0	0	0	0	1	0	0	0	0	0	0	0	0	0	0	−1	−1	−1	1
6	0.34	1	1	0	0	0	0	0	0	1	0	0	0	0	0	0	0	0	0	0	−1	1	1	1
6	−0.22	1	1	0	0	0	0	0	0	1	0	0	0	0	0	0	0	0	0	0	1	−1	1	−1
6	−0.62	1	1	0	0	0	0	0	0	1	0	0	0	0	0	0	0	0	0	0	1	1	−1	−1
6	−0.47	1	1	0	0	0	0	0	0	0	1	0	0	0	0	0	0	0	0	0	−1	−1	−1	1
6	0.33	1	1	0	0	0	0	0	0	0	1	0	0	0	0	0	0	0	0	0	−1	1	1	1
7	−0.23	1	1	0	0	0	0	0	0	0	1	0	0	0	0	0	0	0	0	0	1	−1	1	−1
7	−0.63	1	1	0	0	0	0	0	0	0	1	0	0	0	0	0	0	0	0	0	1	1	−1	−1
7	−0.48	1	1	0	0	0	0	0	0	0	0	1	0	0	0	0	0	0	0	0	−1	−1	−1	1
7	0.32	1	1	0	0	0	0	0	0	0	0	1	0	0	0	0	0	0	0	0	−1	1	1	1
7	−0.24	1	1	0	0	0	0	0	0	0	0	1	0	0	0	0	0	0	0	0	1	−1	1	−1
8	−0.64	1	1	0	0	0	0	0	0	0	0	1	0	0	0	0	0	0	0	0	1	1	−1	−1
8	−0.48	1	1	0	0	0	0	0	−1	−1	−1	−1	0	0	0	0	0	0	0	0	−1	−1	−1	1
8	0.32	1	1	0	0	0	0	0	−1	−1	−1	−1	0	0	0	0	0	0	0	0	−1	1	1	1
8	−0.24	1	1	0	0	0	0	0	−1	−1	−1	−1	0	0	0	0	0	0	0	0	1	−1	1	−1
8	−0.64	1	1	0	0	0	0	0	−1	−1	−1	−1	0	0	0	0	0	0	0	0	1	1	−1	−1

TABLE 3.10B

Lower Half of Design Matrix (Prestige Segment)

		(A) True Partworths																						
		0.00	−0.25	0.25	0.75	0.25	−0.13	−0.38	0.01	0.01	0.00	−0.01	−0.01	−0.01	0.00	0.01	1.59	1.05	−0.39	−1.08	0.07	0.10	0.30	0.25
					Economy Price				Prestige Price				Economy Price				Prestige Price							
Choice Set	Logits (L)	Mean	Cat. (econ)	Cat. (pres)	P1	P2	P3	P4	P1	P2	P3	P4	P1	P2	P3	P4	P1	P2	P3	P4	Eng	M/A	5/6	Int
1	−0.38	1	0	−1	0	0	0	0	0	0	0	0	1	0	0	0	0	0	0	0	−1	−1	1	−1
1	−0.78	1	0	−1	0	0	0	0	0	0	0	0	1	0	0	0	0	0	0	0	−1	1	−1	−1
1	−0.34	1	0	−1	0	0	0	0	0	0	0	0	1	0	0	0	0	0	0	0	1	−1	−1	1
1	0.46	1	0	−1	0	0	0	0	0	0	0	0	1	0	0	0	0	0	0	0	1	1	1	1
1	−0.38	1	0	−1	0	0	0	0	0	0	0	0	0	1	0	0	0	0	0	0	−1	−1	1	−1
2	−0.78	1	0	−1	0	0	0	0	0	0	0	0	0	1	0	0	0	0	0	0	−1	1	−1	−1
2	−0.34	1	0	−1	0	0	0	0	0	0	0	0	0	1	0	0	0	0	0	0	1	−1	−1	1
2	0.46	1	0	−1	0	0	0	0	0	0	0	0	0	1	0	0	0	0	0	0	1	1	1	1
2	−0.37	1	0	−1	0	0	0	0	0	0	0	0	0	0	1	0	0	0	0	0	−1	−1	1	−1
2	−0.77	1	0	−1	0	0	0	0	0	0	0	0	0	0	1	0	0	0	0	0	−1	1	−1	−1
3	−0.33	1	0	−1	0	0	0	0	0	0	0	0	0	0	1	0	0	0	0	0	1	−1	−1	1
3	0.47	1	0	−1	0	0	0	0	0	0	0	0	0	0	1	0	0	0	0	0	1	1	1	1
3	−0.36	1	0	−1	0	0	0	0	0	0	0	0	0	0	0	1	0	0	0	0	−1	−1	1	−1
3	−0.76	1	0	−1	0	0	0	0	0	0	0	0	0	0	0	1	0	0	0	0	−1	1	−1	−1
3	−0.32	1	0	−1	0	0	0	0	0	0	0	0	0	0	0	1	0	0	0	0	1	−1	−1	1
4	0.48	1	0	−1	0	0	0	0	0	0	0	0	0	0	0	1	0	0	0	0	1	1	1	1
4	−0.36	1	0	−1	0	0	0	0	0	0	0	0	−1	−1	−1	−1	0	0	0	0	−1	−1	1	−1
4	−0.76	1	0	−1	0	0	0	0	0	0	0	0	−1	−1	−1	−1	0	0	0	0	−1	1	−1	−1
4	−0.32	1	0	−1	0	0	0	0	0	0	0	0	−1	−1	−1	−1	0	0	0	0	1	−1	−1	1
4	0.48	1	0	−1	0	0	0	0	0	0	0	0	−1	−1	−1	−1	0	0	0	0	1	1	1	1
5	1.62	1	0	1	0	0	0	0	0	0	0	0	0	0	0	0	1	0	0	0	−1	−1	−1	1

5	2.42	1	0	1	0	0	0	0	0	0	0	0	0	0	0	0	1	0	0	0	–1	1	1	1
5	1.86	1	0	1	0	0	0	0	0	0	0	0	0	0	0	0	1	0	0	0	1	–1	1	–1
5	1.46	1	0	1	0	0	0	0	0	0	0	0	0	0	0	0	1	0	0	0	1	1	–1	–1
5	1.08	1	0	1	0	0	0	0	0	0	0	0	0	0	0	0	0	1	0	0	–1	–1	–1	1
6	1.88	1	0	1	0	0	0	0	0	0	0	0	0	0	0	0	0	1	0	0	–1	1	1	1
6	1.32	1	0	1	0	0	0	0	0	0	0	0	0	0	0	0	0	1	0	0	1	–1	1	–1
6	0.92	1	0	1	0	0	0	0	0	0	0	0	0	0	0	0	0	1	0	0	1	1	–1	–1
6	–0.36	1	0	1	0	0	0	0	0	0	0	0	0	0	0	0	0	0	1	0	–1	–1	–1	1
6	0.44	1	0	1	0	0	0	0	0	0	0	0	0	0	0	0	0	0	1	0	–1	1	1	1
7	–0.12	1	0	1	0	0	0	0	0	0	0	0	0	0	0	0	0	0	1	0	1	–1	1	–1
7	–0.52	1	0	1	0	0	0	0	0	0	0	0	0	0	0	0	0	0	1	0	1	1	–1	–1
7	–1.05	1	0	1	0	0	0	0	0	0	0	0	0	0	0	0	0	0	0	1	–1	–1	–1	1
7	–0.25	1	0	1	0	0	0	0	0	0	0	0	0	0	0	0	0	0	0	1	–1	1	1	1
7	–0.81	1	0	1	0	0	0	0	0	0	0	0	0	0	0	0	0	0	0	1	1	–1	1	–1
8	–1.21	1	0	1	0	0	0	0	0	0	0	0	0	0	0	0	0	0	0	1	1	1	–1	–1
8	–1.14	1	0	1	0	0	0	0	0	0	0	0	0	0	0	0	–1	–1	–1	–1	–1	–1	–1	1
8	–0.34	1	0	1	0	0	0	0	0	0	0	0	0	0	0	0	–1	–1	–1	–1	–1	1	1	1
8	–0.90	1	0	1	0	0	0	0	0	0	0	0	0	0	0	0	–1	–1	–1	–1	1	–1	1	–1
8	–1.30	1	0	1	0	0	0	0	0	0	0	0	0	0	0	0	–1	–1	–1	–1	1	1	–1	–1

Cat = Category, (econ) = economy, (pres) = prestige; price levels given in Table 3.9; Eng = Engine, 6 cylinder = –1, 8 cylinder = 1; M/A = Transmission type, M = Manual (–1), A = Automatic (1); 5/6 = Gears, 5 = 5 levels (–1), 6 = 6 levels(1); Int = Interaction between M/A and 5/6 attributes. Levels of Interaction coded as product of M/A and 5/6 coding values.

In this case, it is assumed that there are economy and prestige segments in the population, that is, groups of people who selectively prefer the economy and prestige categories of automobiles. Accordingly, there are separate category effects for the product categories in the economy and prestige segments. That is, the economy segment is assumed to differentially prefer the economy category over the prestige category, and the prestige segment is assumed to differentially prefer the prestige category over the economy category. The source of this differential preference may differ in the two segments. For example, the preference for the economy category in the economy segment might reflect the net disutility of the higher prices characterizing prestige category brands. On the other hand, the preference of the prestige segment for prestige category offerings might reflect a substantially lower disutility for higher prices in the prestige segment, greater utility given to attributes common to prestige brands, greater value attached to the "prestige" provided by brands in the prestige category, or some combination of such considerations.

The different price levels characterizing the economy and prestige categories are accommodated by "nesting" the price attributes within segments. That is, partworths for economy category price levels are separately estimated in the two segments and likewise for prestige price levels. Testing the hypothesis (for example) that the partworths for the economy category partworths are equal in the economy and prestige segments simultaneously tests the hypotheses that the main effect of segment on economy category price and the two-way interaction between segment and economy category prices are zero.

How population segments may be identified is not a topic of this book. In general terms, however, there are two approaches to segmentation, *a priori* and *post hoc*. *A priori* segments are defined in terms of variables that presumably do not depend on the characteristics of sample that will be segmented. In economic, sociology, political science, human geography, and related areas, meaningful *a priori* variables include income, gender, ethnic group, stage of family life cycle and family size, occupational category, political party, area of residence, and so forth. Additional such variables in consumer marketing include product or service usage (amount, package size, and timing); media exposure; distribution channel; and credit usage. Such variables may be defined before a study is undertaken and the number, relative size, and characteristics of segments based on variables of this type are known.

Post hoc segments typically are defined in terms of respondents' similarity (proximity, association, etc.) to other respondents in a sample. For example, psychographic and benefit segments consist of groups of individuals who respond similarly to questions in a survey. In general, the number of segments, their relative size, and their descriptions are found after the fact (Green and Krieger, 1990, p. 4). A variety of approaches have been developed for simultaneous segmentation and estimation using DCE data (see, e.g., the work of DeSarbo, Ramaswamy, and Cohen, 1995; Andrews, Ansari, and Currim, 2002).

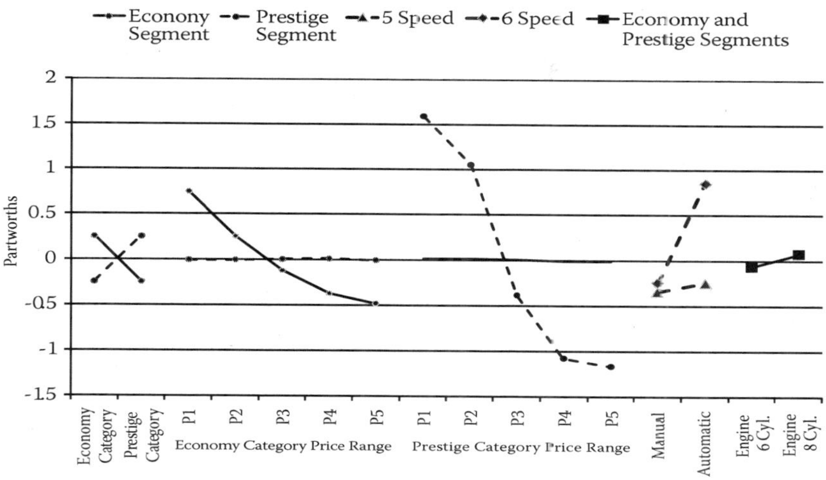

FIGURE 3.1
Partworths as a function of attribute levels (two segments).

Figure 3.1 plots the partworth values of the two segments, with solid lines giving values for the economy segment and dashed lines giving the values of the prestige segment. The three leftmost plots are for the partworths for levels of brand category, price levels of economy category brands, and price levels of prestige category brands. The prestige *segment* has greater utility for the prestige *category* than the economy *category*. The reverse is the case for the economy segment. In other words, there is a category-by-price interaction in both the segments. The key point is that the interpretation of estimated parameters may be misleading when data are the aggregate of subpopulations having differing characteristics.

Furthermore, the nature of these two-way interactions differs in the two segments. The prestige segment is assumed to be price sensitive in the prestige category price range and price insensitive in the economy category price range. Conversely, the economy segment is assumed to be price sensitive in the economy category price range and price insensitive in the prestige category price range. That is, there is a three-way segment-by-category-by-price interaction between these attributes and segment membership.

The next two plots show the partworths for transmission and for number of engine cylinders. There is a two-way interaction between whether the transmission is five or six speed and whether it is manual or automatic. Generally, six-speed transmissions are preferred to five-speed, and automatic is preferred to manual. However, the six-speed automatic transmission is preferred by more than would be expected from the main effects. The segments do not differ in the pattern of preferences for transmissions. The partworths for type of engine do not differ across segments or depend on other attributes.

Clearly, failure to disaggregate the data in this hypothetical example could lead to misleading results. Assuming the segments were of approximately equal size, an aggregate analysis probably would suggest the category effect was zero. Likewise, partworth price estimates in a pooled analysis would be an average of a set of price-sensitive values and a set of price-insensitive values.

3.7 Estimation and Hypothesis Testing

Using the weighted least squares procedures described in Chapter 2, Sections 2.9 and 2.10, regressing the design matrix on the logits shown in Tables 3.10a and b, gives estimated utility values for attribute levels given in at the top of Table 3.10a. The vector of partworths is given by

$$b = (X'S^{-1}X)^{-1} X'S^{-1}\hat{L}_0, \tag{3.3}$$

where $\hat{\mathbf{L}}_0$ is the vector of sample logits, $\mathbf{X}$ is the design matrix, and $\mathbf{S}$ is the estimated variance matrix of logits (see Chapter 2). As noted, the partworth for the \$41,000 price level is given as minus the sum of the partworths for price levels \$21,000 through \$36,000, the partworth for the automatic six-speed transmission is given as minus the sum of the partworths for the other transmissions, and so on. The covariance matrix of the partworths is $\mathbf{S}_b = (\mathbf{X}'\mathbf{S}^{-1}\mathbf{X})^{-1}$.

A test statistic for the model (see Section 2.11) is

$$SS[L_0 = Xb] = \hat{L}_0'S^{-1}\hat{L}_0 - b'(X'S^{-1}X)^{-1}b. \tag{3.4}$$

The statistic is asymptotically a central chi-square distribution with degrees of freedom equal to the number of rows of $\mathbf{X}$ minus the number of columns of $\mathbf{X}$.

As described in Chapter 2, the variance of logits is a function of the proportions from which they are calculated and sample size. The variance may be calculated using the "delta" method and is given in Chapter 2, Section 2.10.

If the choices across sets are independent, the between-set covariances are zero, and the estimated covariance matrix of logits has the following structure:

$$S = \begin{bmatrix} Var(L_{k=1}) & \cdots & 0 & \cdots & 0 \\ \vdots & \vdots & \vdots & \vdots & \vdots \\ 0 & \cdots & Var(L_{k=k'}) & \cdots & 0 \\ \vdots & \vdots & \vdots & \vdots & \vdots \\ 0 & \cdots & 0 & \cdots & Var(L_{k=K}) \end{bmatrix}.$$

where K is the total number of choice sets in the study.

We note that respondents typically are presented choice sets in a predetermined order. No matter how hard they try not to let their current choices be influenced by their previous choices, it is unlikely that they can avoid it. For this reason, a full information estimate of S may be desired.

Figure 3.2 shows upper and lower 95% confidence intervals based on simulated data using partworths shown in Tables 3.10a and 3.10b. The random errors are extreme value given by Equation 1.6. Confidence intervals are based on the mean and standard deviation of 100 replications. Estimates are based on sample sizes of 250.

As discussed in Chapter 2, a hypothesis test on parameters is formulated as

$$SS_H = b'C'(CS_bC')^{-1}Cb. \tag{3.5}$$

The matrix **C** is a known **c × p** matrix defining the hypothesis to be tested; c is the number of parameters—or linear functions of parameters—involved in the hypothesis test; and p is the total number of parameters in the vector **b** (Equation 3.3). This test evaluates the increase in χ^2 that occurs after the parameters, or linear functions of parameters, involved

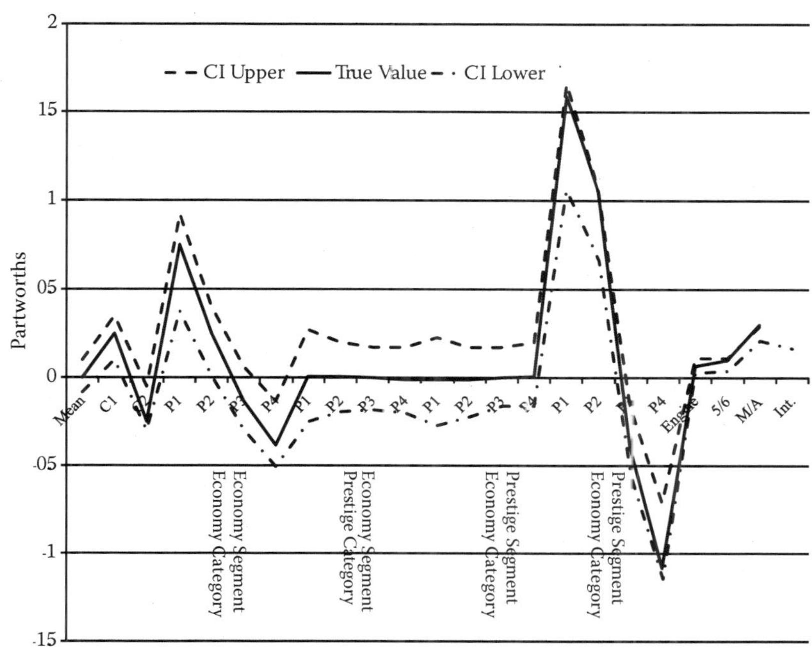

FIGURE 3.2
Confidence intervals for samples of 250 for partworths shown in Table 3.7.

in the hypothesis are restricted to zero. SS_H is distributed approximately χ^2 with degrees of freedom equal to *c*. The appropriate composition of C for a variety of hypotheses can be found in any standard textbook that discusses hypothesis testing within the framework of the general linear model (Morrison, 1967).

Examples of C matrices for eight hypothesis tests for the brand and attributes model are shown in Table 3.11. Matrix **C1** in conjunction with Equation 3.5 tests the hypothesis that the partworth for the economy category in the economy segment is zero. This partworth captures the preference of the economy segment for the economy category independent of other attributes. It may be interpreted as capturing the net of all positive and negative characteristics of offerings in the respective categories. For example, the prestige category has arguably more prestigious brand names and desirable attributes, which for the economy segment are offset by higher prices for offerings in the category.

Matrix **C2** provides a test of the hypothesis that price effects for economy category offerings are simultaneously zero for the economy segment. Matrix **C3** provides a test of the hypothesis that economy segment price effects of prestige category offerings are zero. Based on the values shown in Figure 3.2, the expectation is that the null hypothesis that they are equal to zero would be accepted. Accepting a null hypothesis that the partworths are zero implies that changing the level of the attribute will not change the probability of choosing the profile over choosing a no-choice option.

Matrix **C4** provides a test of the hypothesis that there is no difference in the partworths of the two types of engines. Matrix **C5** tests the hypothesis that there is no difference in partworths for transmission attributes. Based on the values shown in Figure 3.2, the expectation is that the null hypothesis is rejected. Matrix **C6** tests the hypothesis that there is no interaction between transmission attributes.

Matrix **C7** tests the hypothesis that the partworths of economy category price levels are equal in the economy and the prestige segments. This test is equivalent to a test of interaction between economy category price levels and market segment. Matrix **C8** tests the hypothesis that the economy category price partworths are equal to the prestige category partworths in the prestige segment.

The following are results for one of the 100 regression replications using simulated data. The population "true" values for the partworths are shown in the first row labeled (A) of Tables 3.10a and 3.10b. Simulated data for sample sizes of 250 respondents are analyzed. The test of fit for the model (Equation 3.4) is χ^2 = 63.18, degrees of freedom = 57.00, *p* value = .27. The test of lack of fit is not significant, indicating the model fits the data.

Hypothesis sum of squares (Equation 3.5) for hypotheses **C1** through **C8** are as follows:

C1 Economy Segment

Chi square = 348.59 degrees of freedom = 13.00, *p* value = .00.

TABLE 3.11

Contrast Matrices for Eight Hypotheses

	Cat.	Cat.	Economy Price				Prestige Price				Economy Price				Prestige Price							
Mean	**(econ)**	**(pres)**	**P1**	**P2**	**P3**	**P4**	**P1**	**P2**	**P3**	**P4**	**P1**	**P2**	**P3**	**P4**	**P1**	**P2**	**P3**	**P4**	**Eng.**	**M/A**	**5/6**	**Int.**
C1: Part worth for the Economy Category is Zero in the Economy segment																						
0	1	0	0	0	0	0	0	0	0	0	0	0	0	0	0	0	0	0	0	0	0	0
C2: Price Partworths for the Economy Category Are Zero in the Economy Segment																						
0	0	0	1	0	0	0	0	0	0	0	0	0	0	0	0	0	0	0	0	0	0	0
0	0	0	0	1	0	0	0	0	0	0	0	0	0	0	0	0	0	0	0	0	0	0
0	0	0	0	0	1	0	0	0	0	0	0	0	0	0	0	0	0	0	0	0	0	0
0	0	0	0	0	0	1	0	0	0	0	0	0	0	0	0	0	0	0	0	0	0	0
C3: Price Partworths for the Prestige Category Are Zero in the Economy Segment																						
0	0	0	0	0	0	0	1	0	0	0	0	0	0	0	0	0	0	0	0	0	0	0
0	0	0	0	0	0	0	0	1	0	0	0	0	0	0	0	0	0	0	0	0	0	0
0	0	0	0	0	0	0	0	0	1	0	0	0	0	0	0	0	0	0	0	0	0	0
0	0	0	0	0	0	0	0	0	0	1	0	0	0	0	0	0	0	0	0	0	0	0
C4: Engine Partworths Equals Zero																						
0	0	0	0	0	0	0	0	0	0	0	0	0	0	0	0	0	0	0	1	0	0	0

(Continued)

TABLE 3.11 (Continued)

Contrast Matrices for Eight Hypotheses

	Cat.	Cat.	Economy Price				Prestige Price				Economy Price				Prestige Price							
Mean	(econ)	(pres)	P1	P2	P3	P4	P1	P2	P3	P4	P1	P2	P3	P4	P1	P2	P3	P4	Eng.	M/A	5/6	Int.
C5: Transmission Partworths Equal Zero																						
0	0	0	0	0	0	0	0	0	0	0	0	0	0	0	0	0	0	0	0	1	0	0
0	0	0	0	0	0	0	0	0	0	0	0	0	0	0	0	0	0	0	0	0	1	0
0	0	0	0	0	0	0	0	0	0	0	0	0	0	0	0	0	0	0	0	0	0	1
C6: Transmission Interaction between 5/6 Speed and Manual/Automatic Equals Zero																						
0	0	0	0	0	0	0	0	0	0	0	0	0	0	0	0	0	0	0	0	0	0	1
C7: Economy Category Partworths Are Equal in the Economy and the Prestige Segments																						
0	0	0	1	0	0	0	0	0	0	0	–1	0	0	0	0	0	0	0	0	0	0	0
0	0	0	0	1	0	0	0	0	0	0	0	–1	0	0	0	0	0	0	0	0	0	0
0	0	0	0	0	1	0	0	0	0	0	0	0	–1	0	0	0	0	0	0	0	0	0
0	0	0	0	0	0	1	0	0	0	0	0	0	0	–1	0	0	0	0	0	0	0	0
C8: Economy Category Partworths in Economy Segment Are Equal to the Prestige Segment Partworths in the Prestige Segment																						
0	0	0	0	0	0	0	0	0	0	0	1	0	0	0	–1	0	0	0	0	0	0	0
0	0	0	0	0	0	0	0	0	0	0	0	1	0	0	0	–1	0	0	0	0	0	0
0	0	0	0	0	0	0	0	0	0	0	0	0	1	0	0	0	–1	0	0	0	0	0
0	0	0	0	0	0	0	0	0	0	0	0	0	0	1	0	0	0	–1	0	0	0	0

Cat = Category, (econ) = economy, (pres) = prestige; price levels given in Table 3.9; Eng = Engine, 6 cylinder = –1, 8 cylinder = 1; M/A = Transmission type, M = Manual (–1), A = Automatic (1); 5/6 = Gears, 5 = 5 levels (–1), 6 = 6 levels(1); Int = Interaction between M/A and 5/6 attributes. Levels of Interaction coded as product of M/A and 5/6 coding values.

C2 Economy Segment Economy Category

Chi square = 9.82 degrees of freedom = 1.00, *p* value = .00.

C3 Economy Segment Economy Price

Chi square = 19.43 degrees of freedom = 4.00, *p* value = .00.

C4 Economy Segment Prestige Price

Chi square = 0.34 degrees of freedom = 4.00, *p* value = .99.

C5 Engine

Chi square = 15.77 degrees of freedom = 1.00, *p* value = .00.

C6 Transmission

Chi square = 279.55 degrees of freedom = 3.00, *p* value = .00.

C7 Transmission Interaction

Chi square = 83.31 degrees of freedom = 1.00, *p* value = .00.

C8 economy price partworths are equal in economy segment and prestige segment,

Chi square = 9.21 degrees of freedom = 4.00, *p* value = .06.

C9 economy partworths equal the prestige partworths in the prestige segment,

Chi square = 94.17 degrees of freedom = 4.00, *p* value = .00.

In Chapter 4, we illustrate the analysis of data based on Tables 3.10a and 3.10b design and hypothetical values with specific emphasis on the use of orthogonal polynomials for analysis of price parameters.

Appendix 3A: Logit Analysis of Traditional Conjoint Rating Scale Data

The traditional approach to CA is to present respondents with profiles, such as those shown in Table 3.8. Respondents indicate their preference for the profiles using an ordered categorical response scale of purchase intent or other indicator of preference. The common approach for analyzing data of this sort is to assign arbitrary values to the categories. Typically, the integers 1, … , m are assigned, where m is the number of categories. The resulting data typically are analyzed using OLS, although more sophisticated tools may be used.

The approach illustrated addresses two problematical aspects of the traditional approach. First, since in traditional CA the dependent variable typically is response on a rating scale, a second, "choice simulator" stage must be introduced to link predicted preferences with choice proportions (Finkbeiner,

1988). Second, the assignment of integer values to categories implies that the differences between category values are equal.

The approach we illustrate has three elements: a model of the response process, a distribution theory, and an estimation procedure. The model of the response process is based on the "law of categorical judgment," which is patterned on Thurstone's (1927) law of comparative judgment. The law of categorical judgment consists of equations relating stimuli parameters and category boundary parameters to the cumulative proportions each stimulus is judged to be in each response category of a set of categories that are ordered with respect to a given attribute (Torgerson, 1958).

Faced with a scale item, it is assumed that respondents compare their position on the psychological continuum with the positions of the category boundaries of the item, and their response is governed by the differences between the values of these respective parameters. The observed cumulative response distributions are posited to be a function of the cumulative distribution of difference scores on the psychological dimension.

The distribution theory for the model is based on an extension of individual choice theory as developed by McFadden (1974). The extension to aggregate responses for ordered categories is provided by Andrich (1978) and Bechtel and Wiley (1983). An advantage of the present aggregate analysis is that the full covariance matrix of observations is available.

We illustrate the approach using data from a survey of 402 consumers by the Pennsylvania mushroom growers association with the objective to determine which of nine prospective product concepts was most attractive to prospective consumers. Respondents rated the nine concepts on three scales: purchase intent, uniqueness, and believability. This illustrative analysis focuses on one of the scales, the purchase intent scale. This is a five-category

EXHIBIT 3A.1

Illustration of Nine Branding Concepts

Code	Concept
C1	"As few as four fresh mushrooms will provide your daily need for vitamin D."
C2	"Fresh mushrooms on a salad or in a meal are a simple solution for your daily vitamin D needs."
C3	"Fresh mushrooms on a pizza or in pasta are a fun and easy way to add vitamin D to your children's diet."
C4	"Fresh mushrooms can reduce the likelihood of osteoporosis and create a healthy immune system."
C5	"Fresh mushrooms are one of the least-expensive ways to get 100% of your daily vitamin D needs."
C6	"Fresh mushrooms are the only fresh vegetable or fruit that can provide 100% of your daily vitamin D needs."
C7	"Fresh mushrooms provide a low-fat source of vitamin D when compared to milk."
C8	"Fresh mushrooms are a delicious way to get 100% of your daily vitamin D needs."
C9	"Fresh mushrooms are a healthy source for 100% of your daily vitamin D needs."

scale with 5 = very likely to purchase and 1 = very unlikely to purchase. An approach for the analysis of the scales is provided by Landis and Koch (1977). The nine concepts are shown in Exhibit 3A.1.

In this application, the concept of *choice* is not of primary interest. However, we include with the example an illustration of how choice probabilities for the concepts may be calculated using the approach.

The first step of the present procedure is to compute the cumulative response proportions for the sample data. With the coding such that large values are "good," the cumulative proportions are calculated as the proportions below the "highest" category. Table 3A.1 shows the cumulative proportions of respondents using each rating category for each of the nine concepts in Exhibit 3A.1.

Note that there are $m - 1$ category boundaries with an m category scale, for example, the boundary between category 1 and 2, category 2 and 3, category 3 and 4, and so forth. The logits are calculated as $\ln[p/(1 - p)]$, where the ps are the cumulative proportions of responses below the indicated category. For example, for concept 1, 1% of the sample gave category 1 responses (below the boundary between categories 1 and 2), 3% responded in the first *or* second category (below the boundary between category 2 and 3), and so forth. The data are modeled as

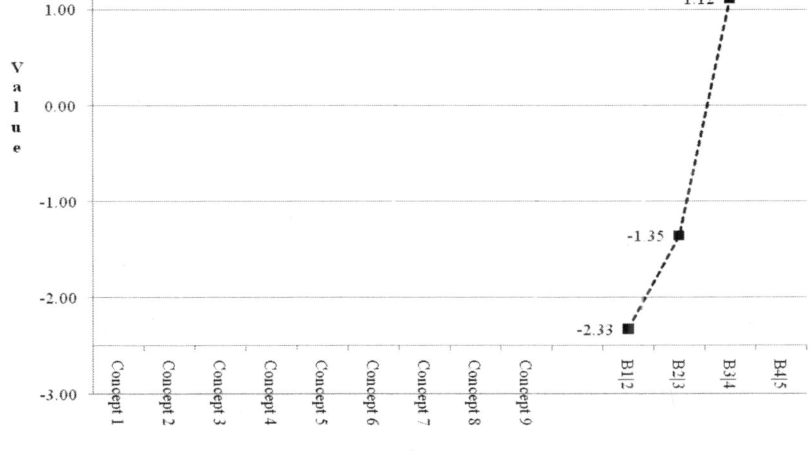

FIGURE 3A.1
Preferences for concepts and boundary values.

TABLE 3A.1

Data for Analysis of Mushroom Concept

Concept	P Below	Cumulative P (n = 401)	Logit	C1	C2	C3	C4	C5	C6	C7	C8	C9	B1	B2	B3
1	1	0.01	−4.19	1	0	0	0	0	0	0	0	0	1	0	0
	2	0.03	−3.48	1	0	0	0	0	0	0	0	0	0	1	0
	3	0.24	−1.17	1	0	0	0	0	0	0	0	0	0	0	1
	4	0.59	0.38	1	0	0	0	0	0	0	0	0	−1	−1	−1
2	1	0.01	0.01	0	1	0	0	0	0	0	0	0	1	0	0
	2	0.02	0.02	0	1	0	0	0	0	0	0	0	0	1	0
	3	0.25	0.25	0	1	0	0	0	0	0	0	0	0	0	1
	4	0.61	0.61	0	1	0	0	0	0	0	0	0	−1	−1	−1
3	1	0.02	−4.03	0	0	1	0	0	0	0	0	0	1	0	0
	2	0.04	−3.18	0	0	1	0	0	0	0	0	0	0	1	0
	3	0.28	−0.92	0	0	1	0	0	0	0	0	0	0	0	1
	4	0.60	0.39	0	0	1	0	0	0	0	0	0	−1	−1	−1
4	1	0.01	−4.60	0	0	0	1	0	0	0	0	0	1	0	0
	2	0.02	−3.67	0	0	0	1	0	0	0	0	0	0	1	0
	3	0.27	−0.97	0	0	0	1	0	0	0	0	0	0	0	1
	4	0.59	0.37	0	0	0	1	0	0	0	0	0	−1	−1	−1
5	1	0.02	−4.03	0	0	0	0	1	0	0	0	0	1	0	0
	2	0.04	−3.12	0	0	0	0	1	0	0	0	0	0	1	0
	3	0.28	−0.92	0	0	0	0	1	0	0	0	0	0	0	1
	4	0.60	0.42	0	0	0	0	1	0	0	0	0	−1	−1	−1
6	1	0.01	−4.60	0	0	0	0	0	1	0	0	0	1	0	0
	2	0.03	−3.32	0	0	0	0	0	1	0	0	0	0	1	0
	3	0.31	−0.80	0	0	0	0	0	1	0	0	0	0	0	1

	4	0.62	0.50	0	0	0	0	0	1	0	0	0	−1	−1	−1
7	1	0.02	−4.03	0	0	0	0	0	0	1	0	0	1	0	0
	2	0.04	−3.06	0	0	0	0	0	0	1	0	0	0	1	0
	3	0.31	−0.82	0	0	0	0	0	0	1	0	0	0	0	1
	4	0.67	0.69	0	0	0	0	0	0	1	0	0	−1	−1	−1
8	1	0.01	−4.60	0	0	0	0	0	0	0	1	0	1	0	0
	2	0.03	−3.32	0	0	0	0	0	0	0	1	0	0	1	0
	3	0.29	−0.91	0	0	0	0	0	0	0	1	0	0	0	1
	4	0.65	0.60	0	0	0	0	0	0	0	1	0	−1	−1	−1
9	1	0.01	−4.89	0	0	0	0	0	0	0	0	1	1	0	0
	2	0.02	−3.77	0	0	0	0	0	0	0	0	1	0	1	0
	3	0.29	−0.91	0	0	0	0	0	0	0	0	1	0	0	1
	4	0.66	0.67	0	0	0	0	0	0	0	0	1	−1	−1	−1

$$u_{ijk} = 1n\{p_{ijk}/(1 - p_{ijk})\} = v_i + \delta_{kj} + \varepsilon_{ijk,} \quad (3A)$$

where

$v_i \equiv$ a fixed (population) preference value for concept i ($i = 1, 9$)
$\delta_j \equiv$ a fixed (population) value for boundary j ($j = 1, 4$), and
$\varepsilon_{ij} \equiv$ random error.

The coding for the design matrix also is shown in Table 3A.1. The concept utility values are estimated as the mean value of the cumulative logits for the respective concepts. Boundary effects are estimated as main-effect *contrasts* about the concept effects, that is, $\Sigma_j \delta_j = 0$. Thus, the value for the boundary between the fourth and fifth categories is estimated as minus the sum of the B1|2, B2|3, and B3|4 boundary values, where, for example, B1|2 refers to the boundary between category 1 and category 2.

For estimation, weighted least squares or full information, maximum likelihood (FIML) may be used. Figure 3A.1 shows preference values for concepts and the values for the boundaries. It is evident that preferences for the concepts do not differ significantly from one another. The values for the boundaries indicate that the two end categories are about half the width of the middle category. That is, the assumption of equal category widths implicit in assigning integer values to boundaries is not supported.

Using the concept values shown in Figure 3A.1, it is evident that respondents primarily used the upper category of the rating scale, and that the concepts do not differ in purchase intent. The proportion of the population that would choose concept 2 from a set consisting of concepts 1, 2, and 3 may be estimated as $p_2 = \exp(v_2)/[\exp(v_1) + \exp(v_2) + \exp(v_3)]$, or $0.39 = e^{2.29}/(e^{2.11} + e^{2.29} + e^{2.22})$.

4

Designs with Ordered Attributes

4.1 Introduction

Discrete choice experiments often, if not usually, have economic attributes; that is, attributes that have ordered levels with associated benefits or costs that monotonically increase with level. Examples include volume, performance measures, and even quality indices. Benefits typically are assumed to increase at a decreasing rate and costs increase at an increasing rate with increasing attribute level. Partworths are assumed to monotonically increase with benefits. Partworths associated with price levels are assumed to monotonically decrease with increasing price level. In the case of individual choice theory that underlies discrete choice experimentation (DCE), it is assumed that there are incremental benefits gained or costs incurred in choosing one profile over another. Decision makers will choose to reduce levels of one attribute to increase levels of another if the incremental gain of switching more than offsets the incremental loss. Clearly, attributes must be ordered for the notions of reducing and increasing attribute levels to make sense.

This chapter discusses two consequences of ordered attributes for the design and analysis of DCEs. When the attribute levels of a DCE are ordered and the levels included may be assumed to be selected from many possible levels, we may describe linear, quadratic, cubic, and higher-order effects of the attribute. The linear effect describes trend (increasing or decreasing) of the dependent variable as the level of the attribute changes. Quadratic effects describe whether the slope changes over the range of the attribute; cubic effects describe whether the slope changes at least twice; and so forth. Section 4.2 describes orthogonal polynomials, which are one way of characterizing the attribute effects of a DCE. We restrict our discussion to linear, quadratic, and cubic effects as these are the ones with the most straightforward economic interpretation; however, quartic and quintic effects are also included in Table 4.1. We also restrict discussion to attributes of 2 through 10 levels. Beyond 10 levels of a single attribute, the number of profiles in a DCE study becomes impractical when there are sufficient attributes to make a study of general interest.

TABLE 4.1

Coefficients of Orthogonal Polynomials

n	**Polynomial**	1	2	3	4	5	6	7	8	9	10
3	Linear	−1	0	1							
	Quadratic	1	−2	1							
4	Linear	−3	−1	1	3						
	Quadratic	1	−1	−1	1						
	Cubic	−1	3	−3	1						
5	Linear	−2	−1	0	1	2					
	Quadratic	2	−1	−2	−1	2					
	Cubic	−1	2	0	−2	1					
	Quartic	1	−4	6	−4	1					
6	Linear	−5	−3	−1	1	3	5				
	Quadratic	5	−1	−4	−4	−1	5				
	Cubic	−5	7	4	−4	−7	5				
	Quartic	1	−3	2	2	−3	1				
7	Linear	−3	−2	−1	0	1	2	3			
	Quadratic	5	0	−3	−4	−3	0	5			
	Cubic	−1	1	1	0	−1	−1	1			
	Quartic	3	−7	1	6	1	−7	3			
8	Linear	−7	−5	−3	−1	1	3	5	7		
	Quadratic	7	1	−3	−5	−5	−3	1	7		
	Cubic	−7	5	7	3	−3	−7	−5	7		
	Quartic	7	−13	−3	9	9	−3	−13	7		
	Quintic	−7	23	−17	−15	15	17	−23	7		
9	Linear	−4	−3	−2	−1	0	1	2	3	4	
	Quadratic	28	7	−8	−17	−20	−17	−8	7	28	
	Cubic	−14	7	13	9	0	−9	−13	−7	14	
	Quartic	14	−21	−11	9	18	9	−11	−21	14	
	Quintic	−4	11	−4	−9	0	9	4	−11	4	
10	Linear	−9	−7	−5	−3	−1	1	3	5	7	9
	Quadratic	6	2	−1	−3	−4	−4	−3	−1	2	6
	Cubic	−42	14	35	31	12	−12	−31	−35	−14	42
	Quartic	18	−22	−17	3	18	18	3	−17	−22	18
	Quintic	−6	14	−1	−11	−6	6	11	1	−14	6

The second consequence of ordered attributes in a DCE is that it is possible for choice sets to have attributes that dominated or are dominated by other attributes in a choice set. A dominating profile is one that has more of every benefit attribute and less of every cost attribute than every other profile in the choice set. A dominated profile is one that has less of every benefit attribute and more of every cost attribute. When such profiles are present in choice sets, respondents need not consider benefits or costs to make their decisions regarding the profile. Including dominating profiles has the effect of reducing the choice set size to the dominating set and no choice. Including dominated profiles has the effect of reducing the choice set size by the size of the dominated set. Section 4.5 discusses Pareto optimal (PO) designs. PO designs are those in which no profile dominates or is dominated by another profile in a choice set.

4.2 Linear, Quadratic, and Cubic Effects

Consider an $s_1 \times s_2 \times \ldots \times s_n$ experiment with attributes $A_1, A_2, \ldots, A_n$, where the factor (or attribute) A_i occurs at s_i levels denoted by $0, 1, 2, \ldots, s_i - 1$. Let each run (or profile) be replicated r times. With a full factorial experiment, there will be $s_2 s_3 \ldots s_n r$ observations with attribute A_1 at the ith level. With a fractional factorial design, there will be the appropriate fraction of this number of observations. In general, let $s_i^{A_1}$ be the number of observations at the ith level of attribute A_1, $s_{ij}^{A_1A_2}$ observations with attribute A_1 at the ith level and attribute A_2 at the jth level and so on.

In Section 2.4, we noted that the main effect of A_1 has $s_1 - 1$ degrees of freedom; the interaction of A_1 and A_2 has $(s_1 - 1)(s_2 - 1)$ degrees of freedom; and so on. In this section, we show that the $s_1 - 1$ degrees of freedom of the main effect may be expressed as $s_1 - 1$ linear and higher-order independent contrasts. In a set of independent contrasts, no contrast can be written as a linear combination of the other contrasts. In the next section, we note that the $(s_1 - 1)(s_2 - 1)$ degrees of freedom of the two-way interactions may be expressed as products of the individual two-factor contrasts. The results in both sections assume that the attribute levels are equally spaced. In Section 4.9, we provide results to obtain orthogonal contrasts with any type of dispersion matrix and equal or unequal spacing of attribute levels. Any pair of orthogonal contrasts has uncorrelated estimates.

When the levels of factor (attribute) A_1 are equally spaced and quantitative, we can form orthogonal contrasts to draw inferences that have meaningful geometric interpretations. Let $T_i^{A_1}$ denote the total response of A_1 with $s_i^{A_1}$ observations, $T_{ij}^{A_1A_2}$ denote the total response with $s_{ij}^{A_1A_2}$ observations, and so on. Clearly,

$$T_i^{A_1} = \sum_{j=0}^{s_2-1} T_{ij}^{A_1A_2}.$$

If we plot the levels i on the x-axis and the response $T_i^{A_1}$ on the y-axis and join the points by a smooth curve, the graph may be linear (straight line with positive or negative slope), quadratic (convex or concave with maximum or minimum), cubic (with two changes in the direction), and so on. Usually, in applied research problems, the graphs are linear or quadratic, showing the optimum level of the attribute in the experimental range. However, cubic effects may also have economic interpretation. For example, suppose the midlevel of a five-level scale is interpreted as a reference point, such as the existing price level of a brand. The two lower levels represent price decreases; the two higher levels represent price increases. As will be illustrated, the cubic effect indicates whether the respondents are equally sensitive to price increases and decreases, more sensitive to price decreases than price increases, or more sensitive to price increases than decreases. In the following, we restrict attention to linear, quadratic, and cubic effects for two-, three-, four-, and five-level attributes.

A linear combination of the responses $T_i^{A_1}$ takes the form $\Sigma_{i=0}^{s_1-1} l_i T_i^{A_1}$. For example,

$$L_1 = T_0^{A_1} - T_{s_1-1}^{A_1};$$

$$L_2 = T_1^{A_1} - T_{s_1-1}^{A_1}; \dots ;$$

$$L_{s_1-1} = T_{s_1-2}^{A_1} - T_{s_1-1}^{A_1}$$

is a set of linear combinations, and

$$L_1^* = T_0^{A_1} - T_1^{A_1};$$

$$L_2^* = T_0^{A_1} + T_1^{A_1} - 2T_2^{A_1}; \dots ;$$

$$L_{s_1-1}^* = T_0^{A_1} + T_1^{A_1} + \dots + T_{s_1-2}^{A_1} - (s_1 - 1)T_{s_1-1}^{A_1}$$

is another set of linear combinations. Clearly, $s_1 - 1$ combinations $L_1, L_2, \dots L_{s_1-1}$ of the effects of attribute A_1 are independent if there do not exist all nonzero values $d_1, d_2 \dots, d_{s_1-1}$ such that $\Sigma_{i=1}^{s_1-1} d_i L_i = 0$. A linear combination is said to be a contrast of the effects of attribute A_1 if $\Sigma l_i = 0$. The set of independent contrasts is not unique, as shown by these two linear combinations.

Two contrasts $\Sigma_{i=0}^{s_1-1} l_i T_i^{A_1}$ and $\Sigma_{i=0}^{s_1-1} m_i T_i^{A_1}$ of the effects of attribute A_1 are said to be orthogonal if $\Sigma_{i=0}^{s_1-1} l_i m_i = 0$. No pair of distinct contrasts $L_1, L_2, \dots L_{s_1-1}$ is orthogonal, whereas each pair of distinct contrasts $L_1^*, L_2^*, \dots L_{s_1-1}^*$ is orthogonal. A set of orthogonal contrasts is independent.

Under the assumption of equal spacing, the linear, quadratic, and cubic contrasts have straightforward geometric interpretations. The following sections illustrate these interpretations for attributes having two ($s_1 = 2$) through five ($s_1 = 5$) levels. Section 4.3 develops interactions between linear and quadratic contrasts. Section 4.4 provides an illustration using orthogonal polynomial contrasts in a DCE experiment.

4.2.1 Attributes at Two Levels

There is a single linear contrast, which is a multiple of

$$A_{1L} = (T_1^{A_1} - T_0^{A_1})$$

that represents the change in response from the lowest to the highest level of attribute A_1. This change may be positive, negative, or zero. When the levels of attribute A_1 are ordered on an interval or ratio scale, the linear contrast A_{1L} may be interpreted as a slope.

4.2.2 Attributes at Three Levels

There are three pairs of contrasts between the three levels. The sum of these is

$$A_{1L} = \{(T_1^{A_1} - T_0^{A_1}) + (T_2^{A_1} - T_0^{A_1}) + (T_2^{A_1} - T_1^{A_1})\}$$

$$= 2(1T_2^{A_1} + 0T_1^{A_1} - 1T_0^{A_1}).$$

There is a single triple of levels, and the difference in the pair of contrasts is

$$A_{1Q} = \{(T_2^{A_1} - T_1^{A_1}) - (T_1^{A_1} - T_0^{A_1})\} = (1T_2^{A_1} - 2T_1^{A_1} + 1T_0^{A_1}).$$

In Table 4.1, we represent this by giving −1, 0, 1 for the linear coefficients and 1, −2, 1 for the quadratic coefficients. The linear effect is a multiple of A_{1L}. The quadratic effect is a multiple of A_{1Q}.

4.2.3 Attributes at Four Levels

There are six pairs of the four levels, and each pair provides a slope parameter. Combining them we get the linear contrast of A_1 denoted by A_{1L}.

$$A_{1L} = \{(T_1^{A_1} - T_0^{A_1}) + (T_2^{A_1} - T_0^{A_1}) + (T_3^{A_1} - T_0^{A_1}) + (T_2^{A_1} - T_1^{A_1}) + (T_3^{A_1} - T_1^{A_1}) + (T_3^{A_1} - T_2^{A_1})\}$$

$$= \{-3T_0^{A_1} - T_1^{A_1} + T_2^{A_1} + 3T_3^{A_1}\}.$$

In Table 4.1, we represent this by giving −3, −1, 1, 3 for the coefficients.

For $s_1 = 4$, there are four triples of levels. The difference in the slopes between any three levels determines the quadratic effect. We combine the four quadratic effects to get the quadratic effect of A_1 denoted by A_{1Q}.

$$A_{1Q} = \{(T_2^{A_1} - T_1^{A_1}) - (T_1^{A_1} - T_0^{A_1})\} + \{(T_3^{A_1} - T_1^{A_1}) - (T_1^{A_1} - T_0^{A_1})\} + \{(T_3^{A_1} - T_2^{A_1}) - (T_2^{A_1} - T_0^{A_1})\} + \{(T_3^{A_1} - T_2^{A_1}) - (T_2^{A_1} - T_1^{A_1})\}$$

$$= 3\{T_0^{A_1} - T_1^{A_1} - T_2^{A_1} + T_3^{A_1}\}.$$

In Table 4.1, we represent this by giving 1, –1, –1, 1 for the coefficients.

Four points determine the cubic effect. The differences between the quadratic effects are the cubic effect. Clearly,

$$A_{1C} = \{T_1^{A_1} - 2T_2^{A_1} + T_3^{A_1}\} - \{T_0^{A_1} - 2T_1^{A_1} + T_2^{A_1}\}$$

$$= \{-T_0^{A_1} + 3T_1^{A_1} - 3T_2^{A_1} + T_3^{A_1}\}$$

We denote this in Table 4.1 by giving –1, 3, –3, 1 for the coefficients.

4.2.4 Attributes at Five Levels

With five levels of factors, there are 10 slopes. Averaging them, we get

$$A_{1L} = \{(T_1^{A_1} - T_0^{A_1}) + (T_2^{A_1} - T_0^{A_1}) + \ldots + (T_4^{A_1} - T_3^{A_1})\}$$

$$= \{-2T_0^{A_1} - T_1^{A_1} + 0T_2^{A_1} + T_3^{A_1} + 2T_4^{A_1}\}.$$

In Table 4.1, we represent this by giving –2, –1, 0, 1, 2 for the coefficients.

There are 10 differences of slopes, and we get the quadratic term as

$$A_{1Q} = \{(T_0^{A_1} - 2T_1^{A_1} + T_2^{A_1}) + (T_0^{A_1} - 2T_1^{A_1} + T_3^{A_1}) + \ldots + (T_2^{A_1} - 2T_3^{A_1} + T_4^{A_1})\}$$

$$= \{2T_0^{A_1} - T_1^{A_1} - 2T_2^{A_1} - T_3^{A_1} + 2T_4^{A_1}\}.$$

There is only one level at which we get the difference of quadratics, and the cubic term is

$$A_{1C} = \{T_4^{A_1} - 2T_3^{A_1} + T_2^{A_1}\} - \{T_2^{A_1} - 2T_1^{A_1} + T_0^{A_1}\}$$

$$= \{-T_0^{A_1} + 2T_1^{A_1} + 0T_2^{A_1} - 2T_3^{A_1} + T_4^{A_1}\}.$$

The linear, quadratic, and cubic effects are multiples of A_{1L}, A_{1Q}, and A_{1C}. Although we have discussed the results for factor (attribute) A_1, the results hold true for any of the n factors (attributes) in the experiment. We give the necessary coefficients in Table 4.1.

4.3 Interaction Components: Linear and Quadratic

A linear combination $\sum_{i,j} d_{ij} T_{ij}^{A_1A_2}$ is called a contrast of the effects of A_1 and A_2 if $\sum_{i,j} d_{ij} = 0$. There are $s_1s_2 - 1$ independent contrasts of the effects of A_1 and A_2, of which $s_1 - 1$ independent contrasts correspond to the effects of A_1 and $s_2 - 1$ independent contrasts correspond to the effects of A_2. The contrasts of the effects of A_1 and A_2 that are orthogonal to the effects of A_1 and A_2 are the interaction effect contrasts, and the number of independent interaction effects of A_1 and A_2 is

$$\{(s_1s_2 - 1) - (s_1 - 1) - (s_2 - 1)\} = (s_1 - 1)(s_2 - 1).$$

This is the interaction degrees of freedom discussed in Section 2.4.

Consider an experiment with equally spaced intervals $s_1 = 3$ and $s_2 = 4$. The linear effect of A_1 is a multiple of $(-T_0^{A_1} + T_2^{A_1})$, and the linear effect of A_2 is a multiple of $(-3T_0^{A_2} - T_1^{A_2} + T_2^{A_2} + 3T_3^{A_2})$. The linear-by-linear interaction effect (linear × linear) is obtained up to a constant term by multiplying the terms as follows:

$$A_{1L}A_{2L} = (-T_0^{A_1} + T_2^{A_1})\ (-3T_0^{A_2} - T_1^{A_2} + T_2^{A_2} + 3T_3^{A_2})$$

$$= 3T_{00}^{A_1A_2} + T_{01}^{A_1A_2} - T_{02}^{A_1A_2} - 3T_{03}^{A_1A_2} - 3T_{20}^{A_1A_2} - T_{21}^{A_1A_2} + T_{22}^{A_1A_2} + 3T_{23}^{A_1A_2}.$$

Next, we consider the linear-by-quadratic interaction effect. The linear effect of A_1 is a multiple of $(-T_0^{A_1} + T_2^{A_1})$, and the quadratic effect of A_2 is a multiple of $(T_0^{A_2} - T_1^{A_2} - T_2^{A_2} + T_3^{A_2})$. The linear-by-quadratic interaction effect (linear × quadratic) is obtained up to a constant term by multiplying the terms as follows:

$$A_{1L}A_{2Q} = (-T_0^{A_1} + T_2^{A_1})\ (T_0^{A_2} - T_1^{A_2} - T_2^{A_2} + T_3^{A_2})$$

$$= -T_{00}^{A_1A_2} + T_{01}^{A_1A_2} + T_{02}^{A_1A_2} - T_{03}^{A_1A_2} + T_{20}^{A_1A_2} - T_{21}^{A_1A_2} - T_{22}^{A_1A_2} + T_{23}^{A_1A_2}$$

$$= \{T_{20}^{A_1A_2} - T_{21}^{A_1A_2} - T_{22}^{A_1A_2} + T_{23}^{A_1A_2}\} - \{T_{00}^{A_1A_2} - T_{01}^{A_1A_2} - T_{02}^{A_1A_2} + T_{03}^{A_1A_2}\}.$$

That is, $A_{1L}A_{2Q}$ is the difference between the quadratic contrast of A_2 at the highest level of A_1 and the quadratic contrast of A_2 at the lowest level of A_1.

Similarly, we have the quadratic-by-linear interaction effect of A_1 and A_2. The quadratic effect of A_1 is a multiple of $(T_0^{A_1} - 2T_1^{A_1} + T_2^{A_1})$, and the linear effect of A_2 is a multiple of $(-3T_0^{A_2} - T_1^{A_2} + T_2^{A_2} + 3T_3^{A_2})$. The quadratic-by-linear interaction effect (quadratic × linear) is obtained up to a constant term by multiplying the terms as follows:

$$A_{1Q}A_{2L} = (T_0^{A_1} - 2T_1^{A_1} + T_2^{A_1})\,(-3T_0^{A_2} - T_1^{A_2} + T_2^{A_2} + 3T_3^{A_2})$$
$$= -3T_{00}^{A_1A_2} - T_{01}^{A_1A_2} + T_{02}^{A_1A_2} + 3T_{03}^{A_1A_2} + 6T_{10}^{A_1A_2} + 2T_{11}^{A_1A_2} - 2T_{12}^{A_1A_2} - 6T_{13}^{A_1A_2}$$
$$-3T_{20}^{A_1A_2} - T_{21}^{A_1A_2} + T_{22}^{A_1A_2} + 3T_{23}^{A_1A_2}.$$

Finally, we have the quadratic-by-quadratic interaction effect of A_1 and A_2. The quadratic effect of A_1 is a multiple of $(T_0^{A_1} - 2T_1^{A_1} + T_2^{A_1})$, and the quadratic effect of A_2 is a multiple of $(T_0^{A_2} - T_1^{A_2} - T_2^{A_2} + T_3^{A_2})$. The quadratic-by-quadratic interaction effect (quadratic × quadratic) is obtained up to a constant term by multiplying the terms as follows:

$$A_{1Q}A_{2Q} = (T_0^{A_1} - 2T_1^{A_1} + T_2^{A_1})\,(T_0^{A_2} - T_1^{A_2} - T_2^{A_2} + T_3^{A_2})$$
$$= T_{00}^{A_1A_2} - T_{01}^{A_1A_2} - T_{02}^{A_1A_2} + T_{03}^{A_1A_2} - 2T_{10}^{A_1A_2} + 2T_{11}^{A_1A_2} + 2T_{12}^{A_1A_2} - 2T_{13}^{A_1A_2}$$
$$+T_{20}^{A_1A_2} - T_{21}^{A_1A_2} - T_{22}^{A_1A_2} + T_{23}^{A_1A_2}$$

Although we have discussed the interaction components for attributes A_1 and A_2, the results hold true for any pair of the n attributes in an experiment. If the linear × linear interaction is significant, the maximum (or minimum) response may be at one of the four vertices of an n-dimensional response surface; if the linear × quadratic or quadratic × linear interaction is significant, the maximum (or minimum) response may lie along the edges of the response surface; and if the quadratic × quadratic interaction is significant, the maximum (or minimum) is at the interior of the response surface.

4.4 An Illustration

Figure 4.1 shows the partworths for price attributes of the illustrative example given in Table 3.10. We wish to describe the shape of the relationship between price levels and partworths. It is evident that the partworths do not change and are approximately zero for the prestige category prices in the

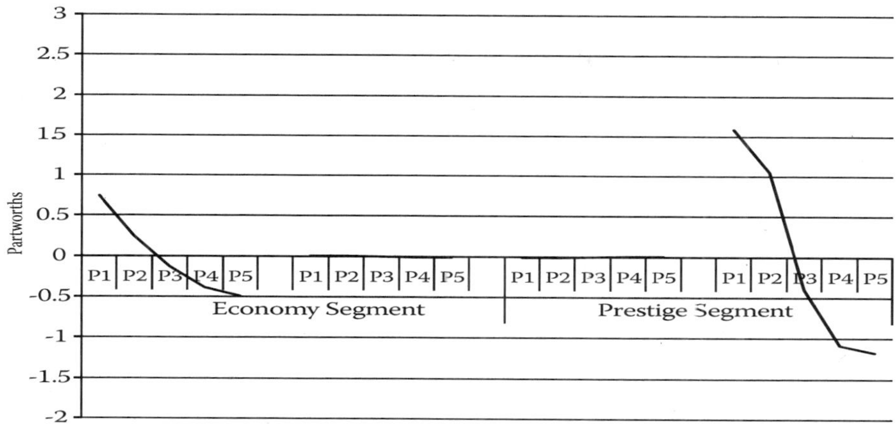

FIGURE 4.1
Relationship between partworths and price levels.

economy segment and for the economy category prices in the prestige segment. The interpretation here is that price changes do not affect the relative proportion choosing the no-choice option in these groups for the respective product categories.

The relationship between partworths and price levels do have distinctive shapes for price levels of economy category cars in the economy segment and for levels of prestige-level cars in the prestige segment. Partworths monotonically increase with decreasing price levels in both groups. However, in the economy segment, partworths for economy category cars appear to increase at an increasing rate as price levels decline. In the prestige segment, the pattern is more complex. As price levels decline toward the midlevel price, partworths appear to increase at an increasing rate. Beyond the midlevel price, however, partworths appear to increase at a decreasing rate.

To determine whether our interpretations of the shapes of the price relationships have statistical support, we decompose the partworth levels into orthogonal polynomial components as described. Table 4.2 presents the coding for orthogonal polynomials for the five-level price attributes shown in Table 3.10.

Figure 4.2 shows the results of a reanalysis of the data described in Section 3.6. The vertical bars present the values of the respective orthogonal polynomial terms. For example, the value of the linear component for the economy category of the economy segment is –.22. The lines show the chi-square value for the respective components against the alternative hypothesis of a zero value. For example, the chi-square value for the linear component value of –.22 is 16.02. All chi-square values are 1 degree of freedom, for which the .95 critical value is 3.80. The value of 16.02 for the linear component is larger than 3.80, so the linear component is significantly different from zero.

TABLE 4.2

Orthogonal and Contrast Coding for Price Levels

	Orthogonal Polynomial Coding for Prices								Contrast Coding for Prices							
	Economy Price				Prestige Price				Economy Price				Prestige Price			
Choice Set	Linear	Quadratic	Cubic	Quartic	Linear	Quadratic	Cubic	Quartic	P1	P2	P3	P4	P1	P2	P3	P4
1	–2	2	–1	1	0	0	0	0	1	0	0	0	0	0	0	0
1	–2	2	–1	1	0	0	0	0	1	0	0	0	0	0	0	0
1	–2	2	–1	1	0	0	0	0	1	0	0	0	0	0	0	0
1	–2	2	–1	1	0	0	0	0	1	0	0	0	0	0	0	0
1	–1	–1	2	–4	0	0	0	0	0	1	0	0	0	0	0	0
2	–1	–1	2	–4	0	0	0	0	0	1	0	0	0	0	0	0
2	–1	–1	2	–4	0	0	0	0	0	1	0	0	0	0	0	0
2	–1	–1	2	–4	0	0	0	0	0	1	0	0	0	0	0	0
2	0	–2	0	6	0	0	0	0	0	0	1	0	0	0	0	0
2	0	–2	0	6	0	0	0	0	0	0	1	0	0	0	0	0
3	0	–2	0	6	0	0	0	0	0	0	1	0	0	0	0	0
3	0	–2	0	6	0	0	0	0	0	0	1	0	0	0	0	0
3	1	–1	–2	–4	0	0	0	0	0	0	0	1	0	0	0	0
3	1	–1	–2	–4	0	0	0	0	0	0	0	1	0	0	0	0
3	1	–1	–2	–4	0	0	0	0	0	0	0	1	0	0	0	0
4	1	–1	–2	–4	0	0	0	0	0	0	0	1	0	0	0	0
4	2	2	1	1	0	0	0	0	–1	–1	–1	–1	0	0	0	0
4	2	2	1	1	0	0	0	0	–1	–1	–1	–1	0	0	0	0
4	2	2	1	1	0	0	0	0	–1	–1	–1	–1	0	0	0	0
4	2	2	1	1	0	0	0	0	–1	–1	–1	–1	0	0	0	0
5	0	0	0	0	–2	2	–1	1	0	0	0	0	1	0	0	0

5	0	0	0	0	–2	2	–1	1	0	0	0	0	1	0	0	0
5	0	0	0	0	–2	2	–1	1	0	0	0	0	1	0	0	0
5	0	0	0	0	–2	2	–1	1	0	0	0	0	1	0	0	0
5	0	0	0	0	–1	–1	2	–4	0	0	0	0	0	1	0	0
6	0	0	0	0	–1	–1	2	–4	0	0	0	0	0	1	0	0
6	0	0	0	0	–1	–1	2	–4	0	0	0	0	0	1	0	0
6	0	0	0	0	–1	–1	2	–4	0	0	0	0	0	1	0	0
6	0	0	0	0	0	–2	0	6	0	0	0	0	0	0	1	0
6	0	0	0	0	0	–2	0	6	0	0	0	0	0	0	1	0
7	0	0	0	0	0	–2	0	6	0	0	0	0	0	0	1	0
7	0	0	0	0	0	–2	0	6	0	0	0	0	0	0	1	0
7	0	0	0	0	1	–1	–2	–4	0	0	0	0	0	0	0	1
7	0	0	0	0	1	–1	–2	–4	0	0	0	0	0	0	0	1
7	0	0	0	0	1	–1	–2	–4	0	0	0	0	0	0	0	1
8	0	0	0	0	1	–1	–2	–4	0	0	0	0	0	0	0	1
8	0	0	0	0	2	2	1	1	0	0	0	0	–1	–1	–1	–1
8	0	0	0	0	2	2	1	1	0	0	0	0	–1	–1	–1	–1
8	0	0	0	0	2	2	1	1	0	0	0	0	–1	–1	–1	–1
8	0	0	0	0	2	2	1	1	0	0	0	0	–1	–1	–1	–1

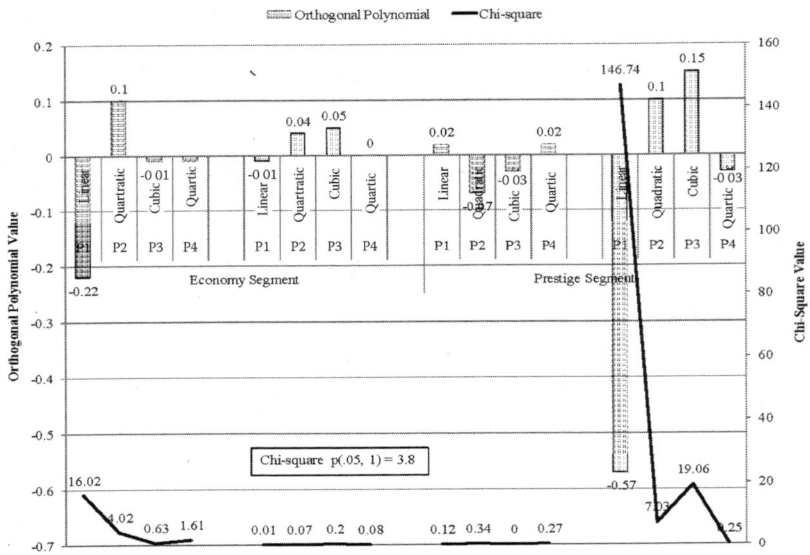

FIGURE 4.2
Orthogonal polynomial components of price partworths.

None of the components for the prestige category in the economy segment or the economy category in the prestige segment differs from zero.

As expected, the linear components for both the economy category in the economy segment and the prestige category in the prestige segment are negative and significantly different from zero. Likewise, the quadratic components are positive and significant in both segments. The positive coefficient indicates the curves are concave to the origin. The cubic component is not significant in the economy segment but is significant in the prestige segment. The lack of significance in the economy segment indicates that the shape of the curve connecting partworths is symmetric about the midlevel price. The significant component for the prestige segment indicates that the curve connecting partworths is not symmetric about the midlevel in the prestige segment. The quartic components are not significant in either segment.

4.5 Pareto Optimal Designs

When profile attributes are ordered, it is possible for profiles to be "better" or "worse" than other profiles on all attributes; that is, a profile will dominate other profiles in a choice set if it offers more of all benefits and incurs less of all costs. As an example, consider the choice of mass transit train options described in terms of four attributes: train fares, train frequency, train routes, and security on trains. We consider only two levels: the current level (0) and an improved level (1) for each attribute, that is, current fares (0) and a 10% reduction in fares (1), current frequency of service (0) or higher frequency (1), current train routes (0) or an increased number of routes (1), and current level of security (0) or improved security (1). If we present the four profile choice sets {1111, 0111, 1011, 1101}, the first profile dominates the others, and the respondent's choice is trivially made. Similarly, if we present a choice set with the profiles {0001, 0111, 1011, 1101}, the first profile is dominated and will not be selected. These examples have in common the fact that respondents need not make trade-offs across attributes when making their choices. If estimated parameters are to be taken as providing information about willingness to make trade-offs, choice sets should be constructed to avoid dominated or dominating alternatives within sets. Such sets are called Pareto optimal (PO) sets.

A study by Wiley, Raghavarao, and Hargreaves (1995), using the employment opportunity attributes, illustrates the practical consequences of having non-PO alternatives in choice sets. Two groups of 55 students were given six sets of four alternatives. The alternatives were described in terms of four attributes: salary, vacation time, training, and an independent index of "best places to work." The attributes were ordered with larger numbers indicating the more desirable level of attribute. A summary of relevant results is provided in Table 4.3. The first group of students was given a mixed set, consisting of two sets that were PO, two sets that had a dominating alternative, and two sets that had a dominated alternative. The sets given to the second group of students were all PO, with the first two sets the same as the PO sets given to group 1.

The results (Wiley, Raghavarao, and Hargreaves, 1995) are the proportions of respondents choosing alternatives in the respective choice sets. Sets 1 and 2 are the same in both groups, and the observed proportions are the same within sampling error (see Table 4.3). For group 1, the third alternative dominates the remaining three in sets 3 and 4. Ninety-six percent of respondents picked the dominating alternative in set 3 and 95% picked the dominating alterative in set 4. For group 1, the third alternative was dominated by each of the remaining three in sets 5 and 6. Nobody chose the dominated alternative in either set. Note that the dominating alternatives in group 1 appeared in group 2 sets 6 and 5, respectively, where they received substantially less than 50% shares. The dominated alternatives in group 1 appeared in group 2 PO sets 4 and 3, respectively. The group 1 set 5 dominated alternative is the most preferred alternative in group 2, set 4, a PO set. The results support the argument that PO choice sets should be used in DCE studies with benefit and cost attributes. The question is, how may PO subsets can be constructed from multiattribute alternatives having ordered attributes?

As with previous material, consider a choice experiment in m attributes, the ith attribute at s_i levels denoted by $0, 1, \ldots, s_i - 1$ for $i = 1, 2, \ldots, m$. Assume that the levels are ordered according to higher benefits and lower costs. Let $(x_1, x_2, \ldots, x_m)$ be a profile where x_i denotes the level of the ith attribute. A

TABLE 4.3

Effect of Dominating and Dominated Alternatives

	Version 1		Version 2		
	Proportion (n = 55)	**Profile (six sets of four)**	**Proportion (n = 55)**	**Profile (six sets of four)**	
Set 1	7.27	1 4 1 1	11.82	1 4 1 1	Set 1
	20.00	1 1 4 1	34.55	1 1 4 1	
	61.82	1 1 1 4	52.73	1 1 1 4	
	10.91	4 1 1 1	10.91	4 1 1 1	
Total	100		100		
Set 2	43.64	2 3 3 3	30.91	2 3 3 3	Set 2
	1.82	3 3 2 3	16.36	3 3 2 3	
	43.64	3 2 3 3	41.82	3 2 3 3	
	10.91	3 3 3 2	10.91	3 3 3 2	
Total	100		100		
Set 3	1.82	1 1 1 3	23.64	1 1 3 1	Set 3
	1.82	3 1 1 1	60.00	3 1 1 1	
Dominating	96.36	4 2 4 4	1.82	1 3 1 1	
	0.00	1 3 1 1	14.55	1 1 1 3	
Total	100		100		
Set 4	1.82	2 2 2 3	5.45	2 2 2 3	Set 4
	3.64	2 2 3 2	38.18	2 2 3 2	
Dominating	94.55	4 3 3 3	47.27	3 2 2 2	
	0.00	2 3 2 2	9.09	2 3 2 2	
Total	100		100		
Set 5	43.64	3 3 3 4	36.36	3 3 3 4	Set 5
	43.64	3 3 4 3	52.73	3 3 4 3	
Dominated	0.00	3 2 2 2	7.27	4 3 3 3	
	12.73	3 4 3 3	3.64	3 4 3 3	
Total	100		100		
Set 6	70.91	2 4 4 4	45.45	2 4 4 4	Set 6
	14.55	4 4 2 4	16.36	4 4 4 2	
Dominated	0.00	1 3 1 1	32.73	4 2 4 4	
	14.55	4 4 2 4	5.45	4 4 2 4	
Total	100		100		

choice set is a subset of *T* distinct profiles of *S*, where *S* is the set of all profiles in an experimental design. Note that the sum of the levels in the respective profiles ranges from 0 to $\Sigma(s_i - 1)$.

Formally, a subset *T* of *S* is said to be a PO subset if for every two distinct profiles $(x_1, x_2, \ldots, x_m)$, $(y_1, y_2, \ldots, y_m) \in T$ there exist subscripts *i* and *j* $(i \neq j)$ such that $x_i < y_i$ and $x_j > y_j$. That is, no profile dominates another in PO sets. Raghavarao and Wiley (1998) note that the choice sets

$$S_l = \left\{ (x_1, x_2, \ldots, x_m) \mid \sum_{i=1}^{m} x_i = l \right\}, l = 0, 1, \ldots, \sum (s - 1)$$

are PO. For the above transportation example, the following three subsets are PO with more than one profile: S_1 = {1000, 0100, 0010, 0001}, S_2 = {1100, 1010, 1001, 0110, 0101, 0011}, and S_3 = {1110, 1101, 1011, 0111}.

Several authors have addressed the problem of constructing PO subsets. Wiley (1978) provides procedures for constructing all PO sets (and the largest PO set in particular) for m^m designs, where *m* is the number of attributes. He also shows that, for this category of design, PO subsets may themselves be organized into balanced subsets, that is, subsets in which all attribute levels appear in the choice set. Krieger and Green (1991) extend that work. They also derive the expected number of dominant entry pairs in a randomly selected choice set in which all attributes have the same number of levels.

A formal treatment of PO designs is provided by Raghavarao and Wiley (1998). Assume that in a DCE experiment there are m $(m \geq 2)$ attributes, $A_1, \ldots, A_m$. Attribute A_i has s_i levels, denoted by $0, 1, \ldots, (s_i - 1)$; $i = 1, \ldots, m$. Raghavarao and Wiley show that the set $S_\alpha = \{(x_1, x_2, \ldots, x_m) \mid \Sigma_{i=1}^{m} x_i = \alpha\}$ is a PO subset for any number attributes and their levels. Note that this result may be used to put profiles into PO sets regardless of how the profiles were constructed. For example, the profiles from a fractional factorial design may be organized into subsets using this approach. The article provides three key results: Let $k = \Sigma_{i=1}^{m} = s_i - m$,

1. The design *D* of attribute profiles resulting from a single PO subset S_l is not a connected main effects plan.
2. For an s^m experiment, the design *D* based on PO subsets $S_{k/2}$ and $S_{(k/2)+1}$ if k is even, and $S_{(k+1)/2}$ and $S_{(k+3)/2}$ if k is odd is a connected main effects plan.
3. In an $s_1 \times s_2 \times \ldots \times s_m$ experiment, $(s_1 \leq s_2 \leq \ldots \leq s_m)$, the main effects plan *D* based on S_l, for $l = s_1 - 1, s_1, s_1 + 1, \ldots, s_m - 1$, is connected.

An important result implicit in the definition of PO subsets is that subsets of PO subsets are themselves PO. In other words, once one has a PO subset, its elements may be further divided into subsets according to any rules or algorithms, including arbitrary ones, and the subsets will be PO.

4.6 Inferences on Main Effects

Let $y_{x_1x_2\ldots x_m}$ be the response corresponding to the profile $(x_1, x_2, \ldots, x_m)$. This response could be the number of points assigned to the profile $(x_1, x_2, \ldots, x_m)$ in the choice set S_l or the logarithm of the ratio of the proportion of respondents choosing $(x_1, x_2, \ldots, x_m)$ to the proportion of respondents choosing the "no-choice" option from S_l. We assume the linear model:

$$E(y_{x_1x_2\ldots x_m}) = \mu + \sum_{i=1}^{m} \alpha_{x_i}^{A_i} + \sum_{\substack{i,j=1\\ i\neq j}}^{m} \alpha_{x_ix_j}^{A_iA_j} + \sum_{\substack{i,j,k=1\\ i\neq j\neq k\neq i}}^{m} \alpha_{x_ix_jx_k}^{A_iA_jA_k} + \ldots + \alpha_{x_1x_2\ldots x_m}^{A_1A_2\ldots A_m},$$

where μ is the general mean, $\alpha_{x_i}^{A_i}$ is the effect of attribute A_i at level x_i, $\alpha_{x_ix_j}^{A_iA_j}$ is the effect of attributes A_i and A_j at levels x_i and x_j, respectively, and so on. The variances and covariances of the responses depend on the nature of the responses.

In this section, we assume the model consisting of the parameters μ and $\alpha_{x_i}^{A_i}$ for $i = 1, 2, \ldots, m$. Consider the problem of estimating the elementary contrasts $\alpha_i^{A_1} - \alpha_j^{A_1}$ of the attribute A_1 for $i < j$ and $i, j = 0, 1, \ldots, s-1$. The pairs of profiles

$$\{(i, x_2^1, \ldots, x_m^1), (i+1, x_2^1, \ldots x_m^1)\},\ \{(i+1, x_2^2, \ldots, x_m^2), (i+2, x_2^2, \ldots x_m^2)\}, \ldots$$
$$\{(j-1, x_2^{j-i}, \ldots, x_m^{j-i}), (j, x_2^{j-i}, \ldots x_m^{j-i})\}$$

occur in choice sets S_l and S_{l+1} where

$$\sum_{u=2}^{m} x_u^t = l - i - t + 1,\ t = 1, 2, \ldots, j-i.$$

It can be verified that the sum of the differences of responses corresponding to these pairs of profiles has expectation $\alpha_i^{A_1} - \alpha_j^{A_1}$, and the contrast $\alpha_i^{A_1} - \alpha_j^{A_1}$ is estimable. Similar results are true for every attribute A_i. Note that S_l must contain profiles like $(s-2, x_2, \ldots, x_m)$, where $x_i \geq 0$ for $i = 2, 3, \ldots, m$, and $(0, x_2, \ldots, x_m)$, where $x_i \leq s-1$ for $i = 2, 3, \ldots, m$. Hence, $s-2 \leq l \leq (m-1)(s-1)$, and we have the following result for an 5^m experiment (Raghowaras and Wiley, 1998)

Theorem 4.1

We can estimate all $m(s-1)$ contrasts of main effects using two consecutive choice sets S_l and S_{l+1} where $s-2 \leq l \leq (m-1)(s-1)$.

The number of profiles in the choice sets S_l and S_{l+1} may be large; therefore, respondents may not be able to choose from the many available profiles or provide an appropriate rating scale. In that case, the experimenter may form nonoverlapping subsets of S_l and S_{l+1} and collect data. The choice of the design is more complicated when the levels of all attributes are not the same. The work of Raghavarao and Wiley (1998) contains more details.

As an illustration, Chitturi, Chitturi, and Raghavarao (2009) assess the influence of hedonic and utilitarian product benefits on customer preference in the context of brand and price information. *Hedonic product benefits* refer to the aesthetics of the product, while *utilitarian product benefits* refer to functionality. To study consumers' relative preference for the hedonic and utilitarian benefits of a product, the authors consider three levels of both the hedonic and utilitarian dimension: low (–1), medium (0), and high (1). Two PO choice sets for estimating the main effects of a 3^2 design are $S_0 = \{-11, 00, 1-1\}$ and $S_1 = \{01, 10\}$. Since these two choice sets have no dominating or dominated profiles, different respondents will make different allocations, and we can gain an insight into the importance of hedonic and utilitarian product benefits. Another example of a conjoint study using unequal set sizes is that of Koelemeijer and Oppewal (1999).

4.7 Inferences on Main Effects in 2^m Experiments

For a 2^m experiment, we can obtain interesting results if we assume the responses are uncorrelated and have the same common variance. Consider any two PO choice sets S_l and S_k with a total number of profiles n. Number the profiles 1, 2, … , n and let $y_1, y_2, \ldots, y_n$ be the average responses or transformed response. We rewrite the main effects model of Section 4.6 as

$$E(y_i) = \mu + \sum_{j=1}^{m} x_{ij}\beta_j,$$

where μ is the general mean, β_j is the main effect of attribute A_j, and x_{ij} is –1 (or 1) depending on the low (or high) level of the attribute A_j in the ith profile. We further assume that the y_i's are uncorrelated and the variance of y_i is σ^2 for every i. After writing the design matrix and some straightforward algebra, it can be shown that the information matrix of m main effects after eliminating μ is an $m \times m$ matrix

$$C = (a_0 - a_2)I_m + \left(a_2 - \frac{a_1^2}{a_0}\right)J_m,$$

where $\mathbf{I}_m$ is an $m \times m$ identity matrix, $\mathbf{J}_m$ is an $m \times m$ matrix of all ones, $a_0 = n$, $a_1 = \sum_{i=1}^{n} x_{i\alpha}$, and $a_2 = \sum_{i=1}^{n} x_{i\alpha}x_{i\beta}$ for any $\alpha, \beta = 1, 2, \ldots m$; $\alpha \neq \beta$.

Raghavarao and Zhang (2002) prove Theorem 4.2 by showing that the C-matrix is nonsingular.

Theorem 4.2

For a 2^m experiment, the design based on any two PO subsets S_l and S_k is a connected main effects plan in which $l \neq k$, $0 < l,k < m$.

While Theorem 4.1 establishes that S_l, S_{l+1} form a connected main effects plan for an s^m experiment, Theorem 4.2 shows that any two distinct PO subsets S_l, S_k form a connected main effects plan for a 2^m experiment.

Since we have several choices for connected main effects plans, we would like to find the plan that is optimal, in some sense, for a 2^m experiment. Note that all these connected main effects plans use different numbers of profiles. Hence, we define information per profile (IPP) ϕ as the reciprocal of the average variance of estimated main effects divided by the number of profiles used in the experiment. Then, ϕ is given by

$$\phi = \frac{m}{a_0 trace(C^{-1})}.$$

The following result was proved by Zhang (2001) and Raghavarao and Zhang (2002):

Theorem 4.3

1. $\phi \leq 1$.
2. $\phi = 1$ for choice sets $S_{\frac{m-t}{2}}$ and $S_{\frac{m+t}{2}}$, when $m = t^2$; and for choice sets $S_{\frac{m-t}{2}}$ and $S_{\frac{m+t+2}{2}}$ (or $S_{\frac{m-t-2}{2}}$ and $S_{\frac{m+t}{2}}$), when $m = t(t + 2)$.

For example, for a 2^9 experiment, the optimal design that maximizes the IPP consists of the two choice sets S_3 and S_6. Here, $m = 9$, $t = 3$, $\frac{m-t}{2} = 3$, and $\frac{m+t}{2} = 6$. The numbers of profiles in the choice sets S_3 and S_6 are $\binom{9}{3} = 84$ and $\binom{9}{6} = 84$, respectively. Clearly, these choice sets are too big and not practicable for application. Therefore, we need to find smaller choice sets with $\phi = 1$. Balanced incomplete block (BIB) designs introduced in Chapter 2 can be used to develop such designs, and these results are discussed in Chapter 5.

4.8 Inferences on Interactions

In Section 4.6, we assumed the linear model

$$E(y_{x_1x_2\ldots x_m}) = \mu + \sum_{i=1}^{m} \alpha_{x_i}^{A_i} + \sum_{\substack{i,j=1 \\ i \neq j}}^{m} \alpha_{x_ix_j}^{A_iA_j} + \sum_{\substack{i,j,k=1 \\ i \neq j \neq k \neq i}}^{m} \alpha_{x_ix_jx_k}^{A_iA_jA_k} + \ldots + \alpha_{x_1x_2\ldots x_m}^{A_1A_2\ldots A_m}$$

where μ is the general mean, $\alpha_{x_i}^{A_i}$ is the effect of attribute A_i at level x_i, $\alpha_{x_i x_j}^{A_i A_j}$ is the effect of attributes A_i and A_j at levels x_i and x_j, respectively, and so on. Raghavarao and Wiley (2006) established the following result for a s^m *experiment*, which provides unbiased estimators for interaction contrasts of the terms $\alpha_{x_i x_j}^{A_i A_j}$ and $\alpha_{x_i x_j x_k}^{A_i A_j A_k}$:

Theorem 4.4

1. We can estimate all $\frac{m(m-1)(s-1)^2}{2}$ contrasts of two-way interactions in an s^m experiment by using PO subsets S_l, S_{l+1}, and S_{l+2}, where $2(s-2) \le l \le (m-2)(s-1)$.
2. We can estimate all $\frac{m(m-1)(m-2)(s-1)^3}{6}$ contrasts of three-way interactions in an s^m experiment by using PO subsets S_l, S_{l+1}, S_{l+2}, and S_{l+3} where $3(s-2) \le l \le (m-3)(s-1)$.

For example, a main effects plan for a 2^4 experiment is

$$S_1 = \{0001, 0010, 0100, 1000\},$$

$$S_2 = \{0011, 0101, 1001, 0110, 1010, 1100\}$$

with

$$S_3 = \{0111, 1011, 1101, 1110\}$$

added to estimate two-way interactions. We are taking $l = 1$ in this example instead of $l = 0$ to avoid single-item choice sets.

The advantage of Theorems 4.1 and 4.4 together is that the experimenter can build a model sequentially by using only two subsets to estimate the main effects and then, if there is significant lack of fit in the model, continue experimentation by collecting more data using S_{l+2} and so on. This aspect of sequential experimentation is described in Chapter 7.

4.9 Orthogonal Polynomials

Let $T_0^A, T_1^A, \ldots, T_{s-1}^A$ be the total responses of a factor A at s levels 0, 1, 2, … , $s - 1$. Let $T' = (T_0^A, T_1^A, \ldots, T_{s-1}^A)$. We can reparameterize as needed so that

$$E(T) = X\alpha, \text{ and } \mathrm{Var}(T) = \sigma^2\Sigma$$

where $\mathbf{X}$ is an $s \times (s - 1)$ matrix of rank $s - 1$, and α is an $(s - 1)^{\text{th}}$-order column vector of parameters. The weighted normal equations estimating α by $\hat{\alpha}$ are

$$(X' \Sigma^{-1} X)\hat{\alpha} = X' \Sigma^{-1} T, \text{ and } Var(\hat{\alpha}) = (X' \Sigma^{-1} X)^{-1} \sigma^2.$$

We want to obtain uncorrelated estimates for suitably transformed parameters as we did with equally spaced levels. Let $\mathbf{X}' \Sigma^{-1} \mathbf{X} = GG'$ be the Cholesky factorization, where G is lower triangular or is based on eigenvalues and vectors of $\mathbf{X}' \Sigma^{-1} \mathbf{X}$. Put $\gamma = G'\alpha$ and $\mathbf{X}^* = \mathbf{X}(G')^{-1}$. Then, $E(T) = \mathbf{X}^*\gamma$, and the weighted normal equations are

$$(X^{*'} \Sigma^{-1} X^*)\hat{\gamma} = X^{*'} \Sigma^{-1} T$$

and

$$Var(\hat{\gamma}) = \left(X^{*'} \Sigma^{-1} X^*\right)^{-1} \sigma^2.$$

Note that

$$X^{*'} \Sigma^{-1} X^* = G^{-1} X' \Sigma^{-1} X (G')^{-1} = G^{-1}(GG')(G')^{-1} = I_{s-1}.$$

Hence, $\hat{\gamma} = G^{-1}\mathbf{X}' \Sigma^{-1} T$, and $Var(\hat{\gamma}) = I_{s-1}\sigma^2$. Such orthogonal estimates can be obtained with unequal spacing of levels and unequal replication of profiles.

We now illustrate the case of orthogonal polynomials with a dispersion matrix that is not a multiple of the identity matrix. Consider a 3^2 experiment in attributes A and B. The resulting nine profiles and no-choice option as a choice set were shown to 100 respondents, and the proportions of respondents choosing the different options were obtained. By summing over the levels of A, we have the following data:

Level of *A*	**Proportion**	**Logit**
0	0.30	ln(6) = 1.7918
1	0.40	ln(8) = 2.0794
2	0.25	ln(5) = 1.6094
No choice	0.05	

Let $\hat{L}_0 = (1.7918, 2.0794, 1.6094)$. The inverse of the estimated dispersion matrix of $\hat{L}_0'$ is

$$S^{-1} = 100\{D(0.30, 0.40, 0.25) - (0.30, 0.40, 0.25)'(0.30, 0.40, 0.25)\}.$$

The linear and quadratic effects of A are given by the model

$$E(\hat{L}_0) = \begin{bmatrix} -1/\sqrt{2} & 1/\sqrt{6} \\ 0 & -2/\sqrt{6} \\ 1/\sqrt{2} & 1/\sqrt{6} \end{bmatrix} \begin{bmatrix} A_L \\ A_Q \end{bmatrix} = X\,\alpha.$$

Now,

$$X'S^{-1}X = \begin{bmatrix} 1 & 4.32 \\ 4.32 & 33 \end{bmatrix} = GG',$$

where $$G = \begin{bmatrix} 1 & 0 \\ 4.32 & 3.7865 \end{bmatrix}.$$

The weighted least squares estimate of A_L and A_Q is

$$\begin{bmatrix} \hat{A}_L \\ \hat{A}_Q \end{bmatrix} = \begin{bmatrix} 1 & 4.32 \\ 4.32 & 33 \end{bmatrix}^{-1} \begin{bmatrix} 54/\sqrt{2} & -2/\sqrt{2} & 56/\sqrt{2} \\ 72/\sqrt{6} & -26/\sqrt{6} & 74/\sqrt{6} \end{bmatrix} \begin{bmatrix} 1.7918 \\ 2.0794 \\ 1.6094 \end{bmatrix}$$

$$\begin{bmatrix} \hat{A}_L \\ \hat{A}_Q \end{bmatrix} = \begin{bmatrix} 273.5172 \\ -33.4054 \end{bmatrix}.$$

The linear and quadratic effects of A with dispersion matrix that is not a multiple of the identity matrix are

$$\begin{bmatrix} \hat{\hat{A}}_L \\ \hat{\hat{A}}_Q \end{bmatrix} = G' \begin{bmatrix} \hat{A}_L \\ \hat{A}_Q \end{bmatrix} = \begin{bmatrix} 129.2058 \\ -126.4895 \end{bmatrix}.$$

4.10 Substitution Rate of Attributes

One may wish to determine how customers are willing to substitute an incremental increase in one attribute for an incremental increase in another attribute. For example, we would like to know how a new graduate is willing to substitute an increase in salary for an increase in the number of vacation days. Consider a 2^3 DCE in three attributes at two levels each. The attributes and levels are salary (0 = \$50,000, 1 = \$55,000); vacation time (0 = 2 weeks, 1 = 3 weeks); and employer's contribution to pension (0 = 5% of salary, 1 = 7% of salary). Two choice sets are formed (i = 1, 2), and data are collected on the choice

TABLE 4.4

2^3 DCE with Artificial Data

Choice Set	Design			Coded Design Matrix				Proportion(p_{ij})	ln(p_{ij}/p_{nc})
	s ($)	*v* (weeks)	*p* (%)	μ	*s*	*v*	*p*		
1	50,000	3	7	1	−0.5	0.5	0.5	0.15	1.0986
1	55,000	2	7	1	0.5	−0.5	0.5	0.30	1.7918
1	55,000	3	5	1	0.5	0.5	−0.5	0.50	2.3025
1	No choice							0.05	
2	55,000	2	5	1	0.5	−0.5	−0.5	0.45	1.5041
2	50,000	3	5	1	−0.5	0.5	−0.5	0.40	1.3863
2	50,000	2	7	1	−0.5	−0.5	0.5	0.05	−0.6931
2	No choice							0.10	

of four profiles (j = 1, 2, 3, and no choice) using 100 new graduates on each of the two choice sets. Artificial data for this scenario are given in Table 4.4, in which salary, vacation time, and employer's contribution to pension are denoted by s, v, and p, respectively. The no-choice option is denoted by nc.

The estimated dispersion matrix is

$$S = \frac{1}{100}\begin{bmatrix} S_1 & O \\ O & S_2 \end{bmatrix}$$

where

$$S_1 = \begin{bmatrix} \frac{1}{.15}+\frac{1}{.05} & \frac{1}{.05} & \frac{1}{.05} \\ \frac{1}{.05} & \frac{1}{.30}+\frac{1}{.05} & \frac{1}{.05} \\ \frac{1}{.05} & \frac{1}{.05} & \frac{1}{.50}+\frac{1}{.05} \end{bmatrix}$$

$$S_2 = \begin{bmatrix} \frac{1}{.45}+\frac{1}{.10} & \frac{1}{.10} & \frac{1}{.10} \\ \frac{1}{.10} & \frac{1}{.40}+\frac{1}{.10} & \frac{1}{.10} \\ \frac{1}{.10} & \frac{1}{.10} & \frac{1}{.05}+\frac{1}{.10} \end{bmatrix}$$

and **O** is a 3 × 3 matrix of zeros. Weighted least squares is used to find the estimates $\hat{s}$ and $\hat{v}$ of the unknown parameters s and v as follows:

$$\begin{bmatrix} \hat{\mu} \\ \hat{s} \\ \hat{v} \\ \hat{p} \end{bmatrix} = \left(X'S^{-1}X\right)^{-1} X'S^{-1}L.$$

Our goal is to find the substitution rate c between an increase in salary s and an increase in the number of vacation days v, that is, determine c so that $s = cv$. If both s and v are nonzero, c can be estimated as s/v, and we can find a confidence interval for c using Fieller's theorem. A brief description of Fieller's theorem is given next; a more detailed description can be found in the work of Finney (1971).

According to Fieller's theorem, if the random variables a and b are thought to be distributed as jointly normal, then for any fixed value r the following probability statement holds if z is an $\alpha/2$ quantile from the standard normal distribution and **V** is the variance-covariance matrix of a and b:

$$\Pr((a - rb)^2 > z^2(V_{aa} - 2r\, V_{ab} + r^2 V_{bb})) = \alpha.$$

Usually, the inequality can be solved for r to yield a confidence interval. For the substitution rate c, the null hypothesis $c = c_0$ is retained against the two-sided alternative if

$$(\hat{s} - c_0\hat{v})^2 \le \chi^2_{1-\alpha}(1)\{\hat{Var}(\hat{s}) + c_0^2\hat{Var}(\hat{v}) - 2c_0 Cov(\hat{s},\hat{v})\} \tag{4.1}$$

where $\chi^2_{1-\alpha}(1)$ is the $(1 - \alpha)$ 100 percentile point of a chi-square distribution with one degree of freedom. Equation 4.1 simplifies to an interval for c_0 if

$$\hat{v}^2 - \chi^2_{1-\alpha}(1)\hat{Var}(\hat{v}) > 0$$

and in that case (4.1) reduces to

$$(c_0 - c_L)(c_0 - c_U) \le 0 \tag{4.2}.$$

We thus get the confidence interval $[c_L, c_U]$ for c with confidence probability $1 - \alpha$.

Weighted least squares is used to estimate the main effects of salary and vacation time as well as their variances and covariance.

We obtain the following estimates:

$$\hat{s} = 1.1983 \quad \hat{Var}(\hat{s}) = 0.3359$$

$$\hat{v} = 0.8628 \quad \hat{Var}(\hat{v}) = 0.3356$$

$$\hat{Cov}(\hat{s}, \hat{v}) = 0.3198.$$

Substituting these estimates in Equation 4.1, we get the 95% confidence interval for c:

$$[-0.2711, 0.9856].$$

In this example C is expected to be nonnegative, hence, the interval can be taken as [0, 0.9856].

5

Reducing Choice Set Sizes

5.1 Introduction

In choice experiments, the investigator organizes profiles into systematically constructed choice sets in which the profiles are designed on the basis of all attributes. When the number of attributes or attribute levels becomes large, the profiles in a single choice set may be too numerous for respondents to make precise decisions. There is a limit to how much information respondents can process without becoming confused or overloaded. For example, consider a 3^5 factorial plan with five factors at three levels each. Using the procedures in Section 4.8, three Pareto optimal (PO) choice sets (S_2, S_3, and S_4) that contain a total of 90 profiles may be used to estimate the main effects and all two-way interactions.

In this chapter, we present strategies for reducing the number of profiles and the sizes of choice sets. One straightforward way to reduce choice set size is to break large choice sets into smaller sets. Section 5.2 illustrates this strategy. A reduced number of profiles and smaller choice set sizes may be generated by subsetting based on overlapping attributes or levels. Sections 5.3 and 5.4 discuss these approaches. A reduced number of profiles and smaller choice set sizes may also be achieved using fractional designs, balanced incomplete block designs (BIBDs), and cyclic designs. Chapters 3 and 4 illustrate the use of fractional designs. Sections 5.5 and 5.6 discuss the use of BIBD and cyclic designs to create designs with reduced numbers of profiles having equal choice set sizes.

Designs in which two-way and higher-order interactions are estimated are particularly prone to requiring many profiles. The number of profiles may be reduced if it is only required that a *subset of interactions* be estimable. Section 5.7 provides designs in which only a subset of two-way interactions is estimated for 2^m and 3^m experiments. Depending on the starting experimental plan, these procedures may be used to create choice sets of the same size across the choice experiment or of varying size.

5.2 Subsetting Choice Sets

The need for subsetting of profiles is immediately encountered when the experimenter is faced with the problem of dividing the runs of a fractional experimental design into sets to conduct discrete choice experimentation (DCE). For example, all the designs illustrated in Chapter 3 were generated as runs of fractional factorial designs. The simplest method of forming smaller sets from large choice sets is to divide each choice set into smaller sets of required size and conduct the study. This is achieved at the cost of an increased number of choice sets.

If the larger choice sets are PO (Chapter 4), the smaller subsets will also be PO. For example, using the methods of Chapter 4 in a 3^4 plan, we can estimate all main effects using two consecutive choice sets S_2 and S_3:

S_2 =	{0002,	0011,	0020,	0101,	0110,
	0200,	1001,	1010,	1100,	2000}
S_3 =	{0012,	0021,	0102,	0111,	
	0120,	0201,	0210,	1002,	
	1011,	1020,	1101,	1110,	
	1200,	2001,	2010,	2100}	

Now, assume we want to use choice sets of size four or less plus a base. We can divide S_2 and S_3 into seven choice sets as follows:

1	{0002,	0011,	0020,		base}
2	{0101,	0110,	0200,	1001,	base}
3	{1010,	1100,	2000,		base}
4	{0012,	0021,	0102,	0111,	base}
5	{0120,	0201,	0210,	1002,	base}
6	{1011,	1020,	1101,	1110,	base}
7	{1200,	2001,	2010,	2100,	base}

In this example, the total number of profiles can be divided into a reasonable number of smaller choice sets. However, as the number of attributes and levels increases, this approach will lead to a large number of choice sets and may not be feasible. Raghavarao and Wiley (2006) give the smallest numbers of profiles for a given number of attributes and levels required to estimate the maximum possible order of interactions.

5.3 Subsetting Levels into Overlapping Sets

In an s^m factorial plan, the s levels $\{0, 1, 2, \ldots, s-1\}$ can be divided into subsets of levels $U_1, U_2, \ldots, U_w$, such that U_i and U_{i+1} are nondisjoint for $i = 1, 2, \ldots,$

$(w - 1)$. We construct PO choice sets for all m attributes using the levels of U_i for each factor $i = 1, 2, \ldots, w$. For example, in a 3^4 experiment, the following three choice sets S_2, S_3, and S_4 with a total of 45 profiles are needed to estimate all main effects and all two-way interactions:

$S_2 =$	{0002,	0011,	0020,	0101,	0110,	
	0200,	1001,	1010,	1100,	2000}	
$S_3 =$	{0012,	0021,	0102,	0111,	0120,	
	0201,	0210,	1002,	1011,	1020,	
	1101,	1110,	1200,	2001,	2010,	2100}
$S_4 =$	{0022,	0112,	0121,	0202,	0211,	
	0220,	1012,	1021,	1102,	1111,	
	1120,	1201,	1210,	2002,	2011,	
	2020,	2101,	2110,	2200}		

Now if a 5^4 experiment is planned, it may be taken as two 3^4 experiments: the first with levels {0, 1, 2} and the second with levels {2, 3, 4}. Thus, the combined design has $2 \times 45 = 90$ profiles instead of 233. In this example, subsets are linked by a single level, but subsets may be linked by the overlap of more than one level. The sets created by subsetting on levels may be interleaved in presentation, so that the respondents will experience all levels going through the presentation order. Also note that in the example we divided the levels into subsets such that the last level overlapped. However, this is not required, and we could also divide the five levels as {0, 1, 3} and {2, 3, 4}. It also is possible to create disjoint subsets that are linked only by the no-choice option that is present in all subsets. However, we recommend having an overlapping level.

With disjoint sets, total lack of fit is the sum of the lack of fit of the respective subsets. When there is overlap, there are two ways of interpreting the design. One interpretation is that subsetting has the effect of creating a block design with the different subsets defining the blocks. When there is an overlap of two or more levels, contrast matrices may be constructed, defining orthogonal polynomials over the range of overlap. For example, with the overlap of two levels, linear components may be estimated over the range of the overlap. A test of equality of the two linear components is a test of equality of slope over the range of overlap in the two blocks. With the overlap of three levels, a block-by-quadratic component may be estimated over the range of the overlap and so forth.

The alternative, response surface design interpretation of the subset design is that one set provides replication/test-of-fit points for the other set. When there is overlap of one or more points, the overlapping points may be interpreted as replication points for estimating error of the other set.

5.4 Subsetting Attributes into Overlapping Sets

In subsetting attributes into overlapping sets, we divide the total number of attributes into subsets $T_1, T_2, \ldots, T_f$ such that T_i and T_{i+1} are nondisjoint for $i = 1, 2, \ldots, (f-1)$. That is, T_i and T_{i+1} are linked by the overlap of at least one attribute. For example, in a 2^7 experiment, the seven attributes $A_1, A_2, \ldots, A_7$ can be divided into two overlapping subsets of four attributes each:

$$T_1 = A_1, A_2, A_3, A_4,$$

$$T_2 = A_4, A_5, A_6, A_7.$$

We form separate main-effect plans based on the attributes in each subset T_i. All main-effect contrasts are estimable, and since T_i and T_{i+1} are nondisjoint, we can test the difference of main effects of any two attributes. A limitation of this approach is that we can only estimate two-factor interactions if both attributes are in the same subset T_i. Therefore, not all two-factor interactions are estimable. Schwabe et al. (2002) and Grossmann et al. (2003) also consider the problem of subsetting by attributes.

In a 2^7 experiment, the following three choice sets S_1, S_2, and S_3 with a total of 63 profiles are needed to estimate all main effects and all two-way interactions:

$S_1 =$	{0000001,	0000010,	0000100,	0001000,	0010000,	0100000,	1000000}
$S_2 =$	{0000011,	0000101,	0000110,	0001001,	0001010,	0001100,	0010001,
	0010010,	0010100,	0011000,	0100001,	0100010,	0100100,	0101000,
	0110000,	1000001,	1000010,	1000100,	1001000,	1010000,	1100000}
$S_3 =$	{0000111,	0001011,	0001101,	0001110,	0010011,	0010101,	0010110,
	0011001,	0011010,	0011100,	0100011,	0100101,	0100110,	0101001,
	0101010,	0101100,	0110001,	0110010,	0110100,	0111000,	1000011,
	1000101,	1000110,	1001001,	1001010,	1001100,	1010001,	1010010,
	1010100,	1011000,	1100001,	1100010,	1100100,	1101000,	1110000}

If we divide the seven attributes of a 2^7 experiment into two overlapping subsets of four attributes each (T_1 and T_2 as discussed), we can use the 2^4 design twice, once with attributes A_1, A_2, A_3, and A_4 and once with attributes A_4, A_5, A_6, and A_7. In a 2^4 experiment, the following three choice sets S_1, S_2, and S_3 with a total of 14 profiles are needed to estimate all main effects and all two-way interactions:

$S_1 =$	{0001,	0010,	0100,	1000}		
$S_2 =$	{0011,	0101,	0110,	1001,	1010,	1100}
$S_3 =$	{0111,	1011,	1101,	1110}		

Thus, the combined design has $2 \times 14 = 28$ profiles (instead of 63). All main effects are estimable, but only those two-factor interactions are estimable for which both attributes are in the same set T_i. For example, the interaction $A_1 A_2$ is estimable, but $A_1 A_7$ is nonestimable. In this example, the subsets T_1 and T_2 are linked by a single attribute A_4, but subsets may be linked by the overlap of more than one attribute. It also is possible to divide the total number of attributes into subsets $T_1, T_2, \ldots, T_f$ such that T_i and T_{i+1} are *disjoint* for $i = 1, 2, \ldots, (f-1)$. In this case, T_i and T_{i+1} may be linked by the overlap of the no-choice option. However, we recommend having overlapping attributes.

When there is no overlap between attributes, the lack of fit for the combined analysis is the sum of the lack of fit for the respective blocks. No separate block effect may be estimated. An overlap of two attributes permits tests of block by the interaction of two overlapping factors.

It is possible to subset by both attributes and levels. Then, for example, the overlap of two levels of two overlapping attributes allows a test of block by bilinear interaction component between the two overlapping attributes over the range of the overlap. An overlap of three levels and three attributes permits tests of bilinear, linear-by-quadratic, and quadratic-by-quadratic interactions in each of the three pairs of overlapping attributes over the range of the attribute overlap. The three-way interaction of blocks, attributes, and levels can also be calculated.

5.5 Designs Generated from a BIBD

Consider a BIBD with parameters $v = m,b,k,r,\lambda$ and its complement BIBD $v' = m,\ b' = b,\ k' = v - k,\ r' = b - r,\ \lambda' = b - 2r + \lambda$. We identify the symbols of the BIBD with m attributes. From the first design, we form a choice set S_k^* of b profiles, where the ith profile is formed by taking the presence (absence) of a symbol as the high (low) level of the corresponding attribute in the ith set of the design. We can similarly form the choice set S_{m-k}^* from the complementary BIBD.

Raghavarao and Zhang (2002) show that for a 2^m design, the information per profile (IPP) for the design based on S_k^* and S_{m-k}^* is the same as the IPP for the design based on S_k and S_{m-k}. The IPP was defined in Theorem 4.2 of Chapter 4. From this result and the existence result of BIBDs (Raghavarao and Padgett, 2005), we get Theorem 5.1:

Theorem 5.1

(a) The design generated from a BIBD with parameters $v = m = t^2, b = t(t+1)$, $k = \frac{t(t-1)}{2},\ r = \frac{t^2-1}{2},\ \lambda = \frac{(t+1)(t-2)}{4}$ and its complement has IPP $\phi = 1$ when $t \equiv 3 \pmod 4$ is a prime or prime power.

TABLE 5.1

Choice Sets Generated Using a BIB Design with Parameters v = 9, b = 12, k = 3, r = 4, and λ = 1.

		S_3^*	S_6^*
Profile	**Design**	**1 2 3 4 5 6 7 8 9**	**1 2 3 4 5 6 7 8 9**
1	A, B, C	1 1 1 0 0 0 0 0 0	0 0 0 1 1 1 1 1 1
2	D, E, F	0 0 0 1 1 1 0 0 0	1 1 1 0 0 0 1 1 1
3	G, H, I	0 0 0 0 0 0 1 1 0	1 1 1 1 1 1 0 0 0
4	A, D, G	1 0 0 1 0 0 1 0 0	0 1 1 0 1 1 0 1 1
5	B, E, H	0 1 0 0 1 0 0 1 0	1 0 1 1 0 1 1 0 1
6	C, F, I	0 0 1 0 0 1 0 0 1	1 1 0 1 1 0 1 0 1
7	A, E, I	1 0 0 0 1 0 0 0 1	0 1 1 1 0 1 1 1 0
8	B, F, G	0 1 0 0 0 1 1 0 0	1 0 1 1 1 0 0 1 1
9	C, D, H	0 0 1 1 0 0 0 0 1	1 1 0 0 1 1 1 1 0
10	C, E, G	0 0 1 0 1 0 1 0 0	1 1 0 1 0 1 0 1 1
11	B, D, I	0 1 0 1 0 0 0 0 1	1 0 1 0 1 1 1 1 0
12	A, F, H	1 0 0 0 0 1 0 1 0	0 1 1 1 1 0 1 0 1

(b) The design generated from a BIBD with parameters $v = m = t^2, b = 2t(t+1)$, $k = \frac{t(t-1)}{2}$, $r = t^2 - 1$, $\lambda = \frac{(t+1)(t-2)}{2}$ and its complement has IPP $\varphi = 1$ when $t \equiv 1 \pmod 4$ is a prime or prime power.

In Theorem 5.1, $t \equiv 3 \pmod 4$ means that $t - 3$ is divisible by 4, and $t \equiv 1 \pmod 4$ means that $t - 1$ is divisible by 4. When $m = 9$, a BIBD exists with parameters $v = 9$, $b = 12$, $k = 3$, $r = 4$, and $\lambda = 1$. Using this design and its complement, we construct choice sets S_3^* and S_6^*, each with only 12 profiles and IPP $\phi = 1$, cutting down the size of the previous 2^9 experiment using S_3 and S_6. The design and the sets S_3^* and S_6^* are given in Table 5.1.

5.6 Cyclic Construction: s Choice Sets of Size s Each for an s^s Experiment

One approach for creating balanced designs based on cyclic rotations of orthogonal Latin square designs is given by Wiley (1978). Next, we describe a cyclic procedure for generating main-effect plans. This strategy may be used to create a main-effect plan in s PO choice sets of size s.

1. Construct the first row as $(0, s-1, 1, s-2, \ldots)$ and cyclically permute to get the first choice set of size s with 0 on the diagonal.
2. Change 0 of the first set to 1 to get the second choice set, 2 to get the third choice set, … and $s-1$ to get the sth choice set.

For example, a main-effect plan for a 5^5 design, in which rows are the choice sets, is as follows:

1	{0 4 1 3 2,	2 0 4 1 3,	3 2 0 4 1,	1 3 2 0 4,	4 1 3 2 0,	Base}
2	{1 4 1 3 2,	2 1 4 1 3,	3 2 1 4 1,	1 3 2 1 4,	4 1 3 2 1,	Base}
3	{2 4 1 3 2,	2 2 4 1 3,	3 2 2 4 1,	1 3 2 2 4,	4 1 3 2 2,	Base}
4	{3 4 1 3 2,	2 3 4 1 3,	3 2 3 4 1,	1 3 2 3 4,	4 1 3 2 3,	Base}
5	{4 4 1 3 2,	2 4 4 1 3,	3 2 4 4 1,	1 3 2 4 4,	4 1 3 2 4,	Base}

Raghavarao and Wiley (2006) also give a cyclic construction strategy to create s^3 choice sets of size at most $s-1$ from which two-way interactions may be estimated.

5.7 Estimating a Subset of Interactions

This section provides strategies for generating smaller designs when only a subset of interactions is of interest. First, we consider situations in which a single two-factor interaction is of interest. A common example is if interest is on one potential two-factor interaction, such as between price and a benefit attribute. For 2^m and 3^m experiments, Raghavarao and Wiley (2006) give connected main-effect plans with smaller choice sets that are capable of estimating *one* two-way interaction as described in Theorems 5.2 and 5.3. The 2^m and 3^m factorial designs are commonly used as screening experiments for which the goal is to identify, among many factors or attributes, the significant few that contribute the most to the response using the smallest number of experimental runs.

Theorem 5.2

(a) The choice sets $S_1^* = \{10...00, 00...01\}$ and $S_2^* = \left\{ x_1 x_2 ... x_m \mid \sum_{i=1}^{m-1} x_i = 1, x_m = 1 \right\}$ are a connected 2^m main-effect plan.

(b) The choice sets S_1^* and $S_2^{**} = S_2^* \cup \{110...00\}$ are a connected main-effect plan and are capable of estimating the two-way interaction A_1A_2.

Theorem 5.3

(a) The choice sets $S_1 = \left\{ x_1x_2...x_m \mid \sum_{i=1}^{m} x_i = 1 \right\}$ and

$$S_2^* = \left\{ x_1x_2...x_m \mid \sum_{i=1}^{m} x_i = 2, x_i = 0 \quad or \quad 2 \right\} \cup \left\{ x_1x_2...x_m \mid \sum_{i=1}^{m-1} x_i = 1, x_m = 1 \right\}$$ are a connected 3^m main-effect plan.

(b) The choice sets $S_3^* = \{110\ldots001, 200\ldots001, 120\ldots000\}$ and $S_4^* = \{220\ldots000, 210\ldots001\}$ along with S_1 and S_2^* of (a) are a main-effect plan that is capable of estimating the two-way interaction A_1A_2.

Chen and Chitturi (2008) consider when subsets of two-way and three-way interactions are of interest. A common example of this situation is when interest is on a subset of interactions inclusive of one factor (e.g., between price and other benefit attributes). Chen and Chitturi give smaller designs based on PO choice sets for estimating main effects and subsets of two-way and three-way interactions inclusive of one factor for 2^m and 3^m plans. Their results are summarized in the following three theorems:

Theorem 5.4
The choice sets $\tilde{S}_1 = \{100\ldots000, 010...000, 000...001\}$ and $\tilde{S}_2 = \left\{ x_1x_2...x_n \mid \sum_{i=1}^{m} x_i = 2, x_1 + x_m \neq 0 \right\}$ are a connected 2^m main-effect plan and are capable of estimating the two-way interactions A_1A_i, $i = 2, 3, \ldots, m$.

As an illustration of Theorem 5.4, consider a 2^5 experiment. If we wish to estimate all main effects and all two-way interactions in a 2^5 experiment, we need a total of 25 profiles in the choice sets S_1, S_2, and S_3. However, if we want to estimate main effects and only those two-way interactions inclusive of one factor, then by Theorem 5.4 only 10 profiles are needed. The choice sets needed to estimate all main effects and two-way interactions involving A_1 are $\tilde{S}_1 = \{10000, 01000, 00001\}$ and $\tilde{S}_2 = \{10001, 01001, 00101, 00011, 11000, 10100, 10010\}$.

Theorem 5.5

The choice sets

$$\bar{S}_1 = \{100\ldots000, 010\ldots000, 0010\ldots000, 000\ldots001\},$$

$$\bar{S}_2 = \left\{ x_1x_2...x_m \mid \sum_{i=2}^{m} x_i = 1, x_1 = 1 \right\} \cup \left\{ x_1x_2...x_m \mid \sum_{i=3}^{m} x_i = 1, x_1 = 0, x_2 = 1 \right\}$$

$$\bigcup\left\{x_1x_2\ldots x_m \mid \sum_{i=4}^{m} x_i = 1, x_1 = x_2 = 0, x_3 = 1\right\}$$

$$\bigcup\left\{x_1x_2\ldots x_m \mid \sum_{i=4}^{m-1} x_i = 1, x_1 = x_2 = x_3 = 0, x_m = 1\right\},$$

$$\overline{S}_3 = \left\{x_1x_2\ldots x_m \mid \sum_{i=3}^{m} x_i = 1, x_1 = x_2 = 1\right\} \bigcup \left\{x_1x_2\ldots x_m \mid \sum_{i=4}^{m} x_i = 1, x_1 = x_3 = 1, x_2 = 0\right\}$$

$$\bigcup\left\{x_1x_2\ldots x_m \mid \sum_{i=4}^{m-1} x_i = 1, x_1 = x_m = 1, x_2 = x_3 = 0\right\}$$

are a connected 2^m main-effect plan and are capable of estimating the two-way interactions A_1A_i, $i = 2, 3, \ldots, m$ and one three-way interaction $A_1A_2A_3$.

As an illustration of Theorem 5.5, consider a 2^6 experiment. If we wish to estimate all main effects, two-way interactions, and three-way interactions in a 2^6 experiment, we need a total of 56 profiles in three choice sets S_1, S_2, and S_3. However, if we want to estimate all main effects, only those two-way interactions inclusive of one factor and one three-way interaction, then by Theorem 5.5, only 26 profiles are needed. The choice sets needed to estimate all main effects, all two-way interactions involving A_1, and $A_1A_2A_3$ in a 2^6 experiment are

$\overline{S}_1$ = {100000, 010000, 001000, 000001}
$\overline{S}_2$ = {110000, 101000, 100100, 100010, 100001, 011000, 010100, 010010, 010001, 001100, 001010, 001001, 000101, 000011}
$\overline{S}_3$ = {111000, 110100, 110010, 110001, 101010, 101001, 100101, 100011}

The symbol $\bigcup$ in the theorem denotes the union of two sets, where the elements of the sets are juxtaposed, deleting any duplicated elements. In this example, $\overline{S}_2$ is the union of four sets as follows:

{11000, 101000, 100100, 100010, 100001} $\bigcup$ {011000, 010100, 010010, 010001} $\bigcup$ {001100, 001010, 001001} $\bigcup$ {000101, 000011}.

By writing all the elements together as one set, we get the set $\overline{S}_2$ with 14 elements.

Theorem 5.6

The choice sets

$\hat{S}_1 = \{100\ldots000, 010\ldots000, 000\ldots001\}$,

$$\hat{S}_2 = \left\{x_1x_2\ldots x_m \mid \sum_{i=1}^{m} x_i = 2, x_i = 0,1, x_1 + x_m \neq 0\right\} \cup (200\ldots000),$$

$$\hat{S}_3 = \left\{x_1x_2\ldots x_m \mid \sum_{i=2}^{m} x_i = 1, x_1 = 2\right\} \cup \left\{x_1x_2\ldots x_m \mid \sum_{i=2}^{m-1} x_i = 2, x_i = 0,2, x_1 + x_m = 1\right\}.$$

$\cup(100\ldots002, 010\ldots002,$

and

$$\hat{S}_4 = \left\{x_1x_2\ldots x_m \mid \sum_{i=2}^{m} x_i = 2, x_i = 0,2, x_1 = 2\right\}$$

are a connected 3^m main-effect plan and are capable of estimating the two-way interactions A_1A_2, $i = 2, 3, \ldots, m$.

As an illustration of Theorem 5.6, consider a 3^5 experiment. If we wish to estimate all main effects and all two-way interactions in a 3^5 experiment, we need a total of 90 profiles. However, if we want to estimate main effects and only those two-way interactions inclusive of one factor, then by Theorem 5.6 only 27 profiles are needed. The choice sets needed to estimate all main effects and two-way interactions involving A_1 are

$\hat{S}_1 =$	{10000,	01000,	00001}	
$\hat{S}_2 =$	{20000,	10001,	10010,	10100,
	11000,	01001,	00101,	00011}
$\hat{S}_3 =$	{20001,	20010,	20100,	21000,
	10002,	10020,	10200,	12000,
	00021,	00201,	02001,	01002}
$\hat{S}_4 =$	{20002,	20020,	20200,	22000}

6

Availability (Cross-Effects) Designs

6.1 Introduction

To this point, the experimental designs discussed have been for situations in which respondents rate a *single* profile or pick a single profile from a choice set. When collecting stated choices using discrete choice experimentation (DCE), however, respondents are given multiple-choice sets of varying size and composition, and it is likely that the composition of the set will influence the attractiveness of the alternatives within the set. When modeling preferences for alternatives and combinations of alternatives selected from choice sets having differing compositions, it sometimes is desirable that the model accommodate the fact that the values of alternatives may be affected by the presence of other alternatives in the choice set. Such effects are referred to as *cross effects.*

Cross effects appear in a variety of forms depending on the nature of the choice task. Availability effects are one form of cross effects. Availability effects may occur in the "pick-one" form of DCE, in which respondents are presented with choice sets and are required to pick one alternative or decline choice. However, suppose an added alternative is identical to one of the original alternatives in a choice set. The expectation would be that the added alternative would divide the share of the identical original and take little share from others in the set. With the multinomial logit model, an added alternative will always take share from the original alternatives in the set in proportion to the shares of the original choice set. For example, imagine a two-alternative choice set in which alternative A receives 50% of the choices, alternative B 25%, and the base option 25%. Suppose an alternative is added to the set that receives 12% of the choices in the expanded choice set. The IIA property (independence from irrelevant alternatives) holds that 6% will be taken from alternative A, 3% from alternative B, and 3% from the base option (see Section 1.4, Chapter 1). A *positive availability effect* occurs when adding an alternative to a choice set results in a probability of choosing another alternative that is *greater* than would be expected if IIA held. A *negative availability effect* occurs when an added alternative results in a probability of choosing another that is *less* than expected if IIA held. To adapt DCE to situations in which IIA may not hold, Batsell and Polking (1985) and Raghavarao and Wiley (1987) propose a generalization of Equation 1.7 to situations in which the value of an alternative is conditional on the composition of the choice set.

Availability designs were specifically developed to allow for situations in which the utility of an alternative depends on the availability of other alternatives in the choice set. In the simplest, "alternatives-only" versions of these models, preference for an alternative is in part an additive function of the other alternatives available in a choice set. Availability effects capture the effect of the availability of an alternative on the attractiveness of other alternatives in choice sets. They may be positive or negative. Let L_{it} be the logit of the proportion of the times alternative i is selected (either singly or in combination in the tth choice set t). The availability model takes the form

$$E(L_{it}) = b_{ii}\delta_{ii} + \sum_{j\neq i}^{j\in t} b_{ij}\delta_{ij}, \tag{6.1}$$

where E(×) stands for the expectation of the random variable in the parenthesis, L_{it} refers to the log of the proportion choosing brand i to the proportion choosing a base option in choice set t, b_{ii} is the "own effect" of brand i, b_{ij} is the cross effect of alternative j on alternative i, and δ_{ij}'s are 0 or 1 elements of a design matrix for an availability model. Conceptually, the own effect of a brand is the partworth of the brand in a choice set consisting of the brand and a base choice option. Equation 6.1 accommodates the possibility that choice set composition will influence the attractiveness of alternatives in a set. For example, the probability of choosing an alternative in a choice set will vary depending on whether there are close substitutes available.

This discussion refers to when respondents choose one component from a choice set. Similar reasoning applies to portfolio choice, that is, the choice of components to form a bundle or portfolio. Much as the availability of other profiles in a choice set may influence the choice of a profile, the presence (or absence) of a component of a bundle may influence the overall attractiveness of the bundle. A key difference between brand choice and portfolio choice (actually portfolio assembly) is that with the latter it is possible for complements to be chosen jointly, and complementary effects may be observed.

A more complicated choice task occurs when respondents may choose more than one alternative in a set. When a choice task may be viewed as assembling a bundle from a set of components, the utilities of bundles will depend on the utilities of components, but the utility of the respective components may be influenced by other components included in the bundle. The probability of including a component in a bundle will depend on whether the component is a complement or a substitute for other available alternatives.

When alternatives are described in terms of attributes, we may also talk of attribute cross effects. In competitive markets, the efforts of a firm and of its competitors influence its market success. The effect of a price reduction will depend on whether competitors match the price cut. More generally, the effect of a change of any attribute of an offering of a firm may depend on whether competitors make a change of their own offering in response.

Attribute cross effects capture the effect of the level of an attribute of one profile on the attractiveness of another profile. As with availability effects, attribute cross effects may be positive or negative and may be of opposite sign. A change in the level of an attribute of profile A may positively affect the attractiveness of profile B, while an equivalent change in the attribute of brand B may negatively affect the attractiveness of brand A. For example, a cross-price attribute effect captures the notion of cross-price elasticity.

In the remainder of this chapter, we first discuss brand-only models, which have only availability effects. We show how Hadamard matrices may be used to create availability designs for brand-only models and discuss some interesting properties of availability models and designs. Following the brands-only section, we turn to portfolio choice. We show how the logit model discussed in Chapter 1 may be generalized to model the assembly of bundles of components or portfolios. Hadamard matrices also may be used to construct designs for estimating portfolio models. An example of a portfolio choice problem and an illustration of the coding of portfolio choice data are provided. Finally, we turn to brands-plus-attribute models. Cross-attribute effects capture the effect of the attribute level of one brand on the value respondents attach to competing brands. An example of a brands-plus-one-attribute model is provided, and several approaches for constructing designs for estimating the parameters of the model are discussed.

6.2 Brands-Only Availability Designs

Raghavarao and Wiley (1987) use 3-designs (doubly balanced incomplete block designs) with parameters $v = 8$, $b = 14$, $r = 7$, $k = 4$, $\lambda_2 = 3$, $\lambda_3 = 1$, introduced by Calvin (1954) to create availability design involving eight brands of soft drinks: Coke, Diet Coke, Pepsi, Diet Pepsi, 7UP, Diet 7UP, Sprite, and Diet Sprite. In their analysis, the cross effects are significant, with the results implying (for example) that if Sprite is replaced by Diet Sprite in a store carrying Sprite, Diet Coke, 7UP, and Pepsi, the sales of Diet Coke will go down more than expected based on the respective own effects of Diet Coke and Sprite.

Other work relevant to brands-only availability designs includes that of Bhaumik (1995), which shows that a 3-design is universally optimal in the class of binary block designs for estimating contrasts of availability effects and cross effects (when random errors are independently and identically distributed [iid] with mean zero and variance σ^2). Raghavarao and Zhou (1998) introduce designs with unequal set sizes called UE 3-designs (see Section 2.3) to estimate all parameters in cross-effects models. Their article proves that the own effects and cross effects may all be estimated *if and only if* the design has unequal choice set sizes, and UE 3-designs are universally optimal in estimating own effects and cross effects. They use UE 3-designs on the eight soft drinks mentioned and show that an optimal four-drink set is {Coke, Diet 7UP, Sprite, and Diet Sprite}.

We note that availability models also provide insights in statistical issues outside the topic of DCE. For example, in Section 6.2.1 we show that cross-effects models may be used to clarify the interpretation of analysis of variance (ANOVA) interactions. Raghavarao, Fédérer, and Schwager (1986) show that cross-effects models resolve some previously unresolved issues of balanced incomplete block designs (BIBDs) with repeated blocks.

The design for an availability model is described by a matrix of 1s and 0s (or 1s and –1s) having n rows and m columns. The columns correspond to the brands in the design, and the rows correspond to the choice sets. A 1 in a column indicates that the brand is present in the choice set. A 0 (or –1) indicates that the brand is absent. The number of 1s in a given row is the choice set size of that choice set. Generally, a 2^n main effect plan and its foldover (see Section 2.7, Chapter 2) will provide an availability design.

With m alternatives for which m is not too large, $(m + 1)$ choice sets with one set of all alternatives and m sets leaving out one alternative at a time is a UE 3-design for estimation of own effects and availability effects. With $m = 4$ and brands {A, B, C, D}, the following is the design:

{A, B, C, D}
{A, B, C}
{A, B, D}
{A, C, D}
{B, C, D}

D. A. Anderson and Wiley (1992) illustrate that availability designs for m alternatives can be constructed by taking any m columns of a Hadamard H_{4t} and its foldover, that is,

$$\begin{bmatrix} H_{4t} \\ -H_{4t} \end{bmatrix}, \tag{6.2}$$

where $4t \geq m$. When H_{4t} has the first row of all 1s, the $(4t + 1)$th row has all –1s, and it corresponds to an empty choice set. Thus, the design has $8t - 1$ choice sets, where at least one choice set has all m alternatives and the remaining choice sets do not have all alternatives. That is, the choice sets are of unequal set size.

Table 6.1 illustrates the approach. Matrix **Av-5** (Panel B) is generated by folding over the Hadamard matrix of order eight, H_8 (Panel A), and dropping the first three columns. The first choice set of **Av-5** contains all of the

TABLE 6.1

Availability Designs Created from Hadamard Matrices

(A) H_8								(B) Availability Design Av-5					nc	Set Size
1	1	1	1	1	1	1	1	1	1	1	1	1	1	5
1	−1	1	−1	1	−1	1	−1	−1	1	−1	1	−1	1	2
1	1	−1	−1	1	1	−1	−1	−1	1	1	−1	−1	1	2
1	−1	−1	1	1	−1	−1	1	1	1	−1	−1	1	1	3
1	1	1	1	−1	−1	−1	−1	1	−1	−1	−1	−1	1	1
1	−1	1	−1	−1	1	−1	1	−1	−1	1	−1	1	1	2
1	1	−1	−1	−1	−1	1	1	−1	−1	−1	1	1	1	2
1	−1	−1	1	−1	1	1	−1	1	−1	1	1	−1	1	3
								−1	−1	−1	−1	−1	1	0
								1	−1	1	−1	1	1	3
								1	−1	−1	1	1	1	3
								−1	−1	1	1	−1	1	2
								−1	1	1	1	1	1	4
								1	1	−1	1	−1	1	3
								1	1	1	−1	−1	1	3
								−1	1	−1	−1	1	1	2

(C) H_{8a}								(D) Availability Design Av-5a					nc	Set Size
1	1	1	1	1	1	1	−1	1	1	1	1	−1	1	4
1	−1	1	−1	1	−1	1	1	−1	1	−1	1	1	1	3
1	1	−1	−1	1	1	−1	1	−1	1	1	−1	1	1	3
1	−1	−1	1	1	−1	−1	−1	1	1	−1	−1	−1	1	2
1	1	1	1	−1	−1	−1	1	1	−1	−1	−1	1	1	2
1	−1	1	−1	−1	1	−1	−1	−1	−1	1	−1	−1	1	1
1	1	−1	−1	−1	−1	1	−1	−1	−1	−1	1	−1	1	1
1	−1	−1	1	−1	1	1	1	1	−1	1	1	1	1	4
								−1	−1	−1	−1	1	1	1
								1	−1	1	−1	−1	1	2
								1	−1	−1	1	−1	1	2
								−1	−1	1	1	1	1	3
								−1	1	1	1	−1	1	3
								1	1	−1	1	1	1	4
								1	1	1	−1	1	1	4
								−1	1	−1	−1	−1	1	1

H_{8a} created by multiplying the last column of H_8 by −1

alternatives. The second choice set, corresponding to the second row of H_8, contains two alternatives corresponding to the fifth and seventh columns of H_8. The availability design **Av-5** has 1 set with five alternatives, 1 with four, 6 with three, 6 with two, 1 with one, and 1 empty, for a total of 15 choice sets. The first two rows of this design provide the coding for Equations 6.3a and 6.3b.

An alternative design **Av-5a** (Table 6.1, Panel D) is generated by multiplying the last column of H_8 by –1 (Table 6.1, Panel C) and folding over this alternative design. The first choice set of **Av-5a** has all alternatives except the last. The second choice set contains three alternatives corresponding to the fifth, seventh, and eighth columns of H_{8a}. Design **Av-5a** has 4 choice sets with four alternatives, 4 with three alternatives, 4 with two alternatives, and 4 with one alternative for a total of 16 choice sets.

To illustrate the use of a Hadamard matrix to create an alternative-only model, imagine a five-alternative design, one in which the first two choice sets have the composition $\{i = 1, 2, 3, 4, 5; t = 1\}$ and $\{i = 2, 4; t = 2\}$, given by the first two rows of **Av-5**, Table 6.1, Panel B. The cross-effect models for the value L_{it} of alternative $(i = 1)$ in set $(t = 1)$ and alternative $(i = 2)$ in choice set $(t = 2)$ are, respectively,

$$Y_{11} = \mathrm{E}(\mathrm{L}_{11}) = +b_{11} + b_{12} + b_{13} + b_{14} + b_{15} - b_{00} \tag{6.3a}$$

$$Y_{22} = E(L_{22}) = -b_{21} + b_{22} - b_{23} + b_{24} - b_{25} - b_{00} \tag{6.3b}$$

where b_{ij}, $i = j$ are the own effects and b_{ij}, $i \neq j$ are the cross effects. Note that the partworths estimated for alternatives are actually the contrast between the value of the alternative and the value of the base option. When the base option is no choice, the value of b_{00} conventionally is taken to equal zero.

The choice proportions that would prevail if IIA held provide a benchmark that defines equal substitutability between brands and cross effects pickup deviations from equal substitutability. A positive cross effect implies that the share received by an alternative in the presence of the other alternative is greater than expected given its partworth in the absence of the alternative. A negative partworth implies the share is less than expected.

To understand why marketing managers and policy makers are interested in brands-only cross-effects models, note that the m^2 own and cross effects of an m alternative experiment may be organized as

$$V = \begin{bmatrix} b_{11} & b_{12}\cdots & b_{1m} \\ b_{21} & b_{22}\cdots & b_{23} \\ \vdots & \vdots & \vdots \\ b_{m1} & b_{m2} & b_{m3} \end{bmatrix}. \tag{6.4}$$

The parameters along the main diagonal are the own effects. The parameters in the row i (excluding b_{ii}) are called *vulnerabilities*. They capture the effects of the availability of competing brands on brand i. The parameters in column j (excluding b_{jj}) are brand j's "clout." They capture the effect of the availability of brand j on competing brands. As noted, the elements of the matrix given in Equation 6.4 may have the same or opposite signs. They are not necessarily symmetric ($b_{ij} = b_{ji}$) or antisymmetric ($b_{ij} = -b_{ji}$).

From a result of Sprott (1954), it is known that, omitting the first row of H_{4t} and $-H_{4t}$ of display Equation 6.2, the resulting choice sets form a 3-design (see the work of Raghavarao and Padgett, 2005) in $4t$ alternatives, $8t - 2$ sets, where each symbol occurs in $4t - 1$ sets, every pair of distinct alternatives occurs in $\lambda_2 = 2t - 1$ sets, and every triplet of distinct alternatives occurs in $\lambda_3 = t - 1$ sets. To this, we are augmenting a choice set of all alternatives to get a 3-design with unequal set sizes (see Raghavarao and Zhou, 1998). Since deleting alternatives from a 3-design still results in a 3-design, the availability designs given by D. A. Anderson and Wiley (1992) are 3-designs with unequal set sizes.

Raghavarao and Zhou (1998) prove that 3-designs with unequal set sizes are universally optimal when Y_{ij}'s are uncorrelated with equal variances. From that result, given the number of alternatives and the number of choice sets and their sizes, the designs constructed by dropping columns from Hadamard matrices (Equation 6.2) are universal optimal for a given $4t \geq m$.

Since availability designs may be constructed for every t satisfying $4t \geq m$ using Hadamard matrices, the behavior of such designs for every t is of interest. Raghavarao and Wiley (2008) show that there are two surprising properties of these designs. First, while the number of choice sets and observations will increase with t (resulting in smaller variances for the estimated parameters), the information per profile (IPP) of the designs is the same regardless of t. When the covariance matrix is iid, the IPP is $1/(5m - 4)$. Note that the information is the reciprocal of the average variance of the parameters.

Second, some of the choice sets may be duplicated. Deleting the duplicate choice sets, all m^2 parameters from an availability model may still be estimated. The structure of the matrix makes a difference in this case, and the information per response depends on the order and columns chosen of the Hadamard matrix. However, in the range of Hadamard matrices considered relevant for research on availability effects (say H_8 to H_{20}, with 4 through 20 alternatives), for $m > 8$ there are no duplicate choice sets. For $4 \leq m \leq 8$, the IPP is the same for designs with duplicates and those without duplicates, with two exceptions. One exception is for designs for $m = 4$ constructed from H_{20}. The other exception is for the designs for $m = 5$ constructed from H_{12}.

6.2.1 Relationships between Availability Effects and Interactions

The standard ANOVA model with main effects and interactions provides an alternative model for analyzing response in competitive environments. Both the ANOVA and availability models allow for the response of an alternative to depend on the presence of other available alternatives. Both availability effects and interactions may be positive or negative. There are m own effects, $m(m - 1)$ first-order cross effects, and $m(m - 1)(m - 2)/2$ second-order availability effects. There are m main effects, $m(m - 1)/2$ two-way interactions, and

$m(m-1)(m-2)/6$ three-way interactions. The fact that there are precisely twice as many first-order cross effects as two-way interactions and thrice as many second-order cross effects as three-way interactions suggests that there may be a straightforward relationship between them; in fact, there is.

Availability effects and interactions are not identical, however; interactions are symmetric (the interaction between A and B is the same as between B and A); availability effects are not necessarily symmetric (the availability effect of A on B may not even have the same sign as that of B on A).

There are complicating issues stemming from the differing treatments in an ANOVA analysis and an availability model analysis of the dependent variables, Y_{ij} or $Y_j = \Sigma_i Y_{ij}$. In a field experiment, the dependent variable in an ANOVA analysis Y_j would be the *total* sales from a store. In an availability model analysis, it would be the sales of individual brands i in stores j, Y_{ij}. These measures might be approximated in a conjoint analysis (CA) experiment by presenting respondents with sets of brands and getting a preference measure for the respective sets for an ANOVA analysis versus getting individual preference measures for an availability model analysis.

The dependent variable for DCE frequently, if not usually, is proportion of sales or choice. As described in other chapters, the availability analysis is then based on the logit of the within-choice-set share L_{ij} of each brand. The comparable dependent variable for an ANOVA analysis is the logit L_j of the share of respondents who make "some choice" from the choice set to those making "no choice."

In what follows, we assume a second-order availability effects model for the response Y_{ij}, and we model Y_j using main effects, two-factor, and three-factor interactions without a general mean. In this context, we are using parentheses for cross effects in the subscripts. Under this setting from Table 6.2, we have

$E(Y_{A1}) = \beta_{AA}$, $E(Y_{A4}) = \beta_{AA} + \beta_{A(B)}$, $E(Yy_{A5}) = \beta_{AA} + \beta_{A(C)}$, $E(Y_{A7}) = \beta_{AA} + \beta_{A(B)} + \beta_{A(C)} + \beta_{A(BC)}$, etc.,

where β_{AA} is the brand's own effect; $\beta_{A(B)}$ and $\beta_{A(C)}$ are first-order cross effects of brands B and C on A, respectively; and $\beta_{A(BC)}$ is the second-order cross effect of the BC combination. As noted, the own effect is the effect on sales of the brand when it is in a choice set consisting of itself and a no-choice option. First-order availability effects are the differential effects of adding each of the other brands to this single-brand choice set. The first-order availability effects of $\beta_{A(B)}$ and $\beta_{B(A)}$ are not necessarily equal and in fact may be of opposite sign. Second-order availability effects are the differential effects of the first-order interactions between $\beta_{A(B)}$ and $\beta_{A(C)}$ on the brand A own effect.

Raghavarao and Wiley (2009b) provide the following relationships between availability effects and interaction effects of an ANOVA study: Let Y_j be the response from choice set j, $j = 1, 2, \ldots, 7$. The choice sets and responses are given in Table 6.2. The responses Y_j are aggregate responses for the choice set, such as sales of all the brands in a category, for example, the choice set

TABLE 6.2

Responses on Brands and Choice Sets

Choice Set	Brand A	Brand B	Brand C	Choice Set Response
1	Y_{A1}			Y_1
2		Y_{B2}		Y_2
3			Y_{C3}	Y_3
4	Y_{A4}	Y_{B4}		Y_4
5	Y_{A5}		Y_{C5}	Y_5
6		Y_{B6}	Y_{C6}	Y_6
7	Y_{A7}	Y_{B7}	Y_{C7}	Y_7

response is the sum of the sales of the brands in the respective choice sets: $Y_1 = Y_{A1}, \ldots, Y_4 = Y_{A4} + Y_{B4}, \ldots, Y_7 = Y_{A7} + Y_{B7} + Y_{C7}$. Estimated effects of an ANOVA model are based on the values of Y_i.

We assume a second-order availability effects model for the response Y_{ij}. We model Y_j using main effects, two-factor, and three-factor interactions without a general mean, and we discussed the availability effects model in this section.

There are 12 equations in 12 unknowns and the best, unbiased estimators of the parameters are

$$\hat{\beta}_{A(BC)} = Y_{A1} + Y_{A7} - Y_{A4} - Y_{A5},$$

$$\hat{\beta}_{A(B)} = Y_{A4} - Y_{A1}, \hat{\beta}_{A(C)} = Y_{A5} - Y_{A1},$$

$$\hat{\beta}_{AA} = Y_{A1}.$$

Defining other brand effects and availability effects parameters and estimating them, we get the following result:

$$\begin{aligned} &\hat{\beta}_{A(BC)} + \hat{\beta}_{B(AC)} + \hat{\beta}_{C(AB)} \\ &= Y_{A1} + Y_{A7} - Y_{A4} - Y_{A5} + Y_{B2} + Y_{B7} - Y_{B4} - Y_{B6} + Y_{C3} + Y_{C7} - Y_{C5} - Y_{C6} \\ &= Y_1 + Y_2 + Y_3 + Y_7 - Y_4 - Y_5 - Y_6 \end{aligned} \tag{6.4}$$

$$\hat{\beta}_{A(B)} + \hat{\beta}_{B(A)} = Y_{A4} - Y_{A1} + Y_{B4} - Y_{B2} = Y_4 - Y_1 - Y_2 \tag{6.5}$$

We also have

$$E(Y_1) = \alpha_A, E(Y_2) = \alpha_B, E(Y_3) = \alpha_C,$$

$$E(Y_4) = \alpha_A + \alpha_B + \alpha_{AB},\ E(Y_5) = \alpha_A + \alpha_C + \alpha_{AC},\ E(Y_6) = \alpha_B + \alpha_C + \alpha_{BC},$$

$$E(Y_7) = \alpha_A + \alpha_B + \alpha_C + \alpha_{AB} + \alpha_{AC} + \alpha_{BC} + \alpha_{ABC},$$

where α_i is the main effect of factor i; α_{ij} is the two-factor interaction of factors i and j; and α_{ijk} is the three-factor interaction of factors i, j, and k; $i \neq j \neq k$; $i, j, k =$ A, B, C. Again, there are seven equations in seven unknowns, and we get the best unbiased estimators

$$\hat{\alpha}_{ABC} = Y_1 + Y_2 + Y_3 + Y_7 - Y_4 - Y_5 - Y_6 \tag{6.6}$$

$$\hat{\alpha}_{AB} = Y_4 - Y_1 - Y_2. \tag{6.7}$$

From Equations 6.4 and 6.6, we have

$$\hat{\alpha}_{ABC} = \hat{\beta}_{A(BC)} + \hat{\beta}_{B(AC)} + \hat{\beta}_{C(AB)},$$

and from Equations 6.5 and 6.7, we have

$$\hat{\alpha}_{AB} = \hat{\beta}_{A(B)} + \hat{\beta}_{B(A)}.$$

Now, suppose that there are no second-order availability effects; that is, we drop the parameters $\beta_{A(BC)}$, $\beta_{B(AC)}$, and $\beta_{C(AB)}$ and assume the availability effects model:

$$E(Y_{A1}) = \beta_{AA},\ E(Y_{A4}) = \beta_{AA} + \beta_{A(B)},\ E(Y_{A5}) = \beta_{AA} + \beta_{A(C)},\ E(Y_{A7}) = \beta_{AA} + \beta_{A(B)} + \beta_{A(C)}.$$

In this case, $Y_{A4} - Y_{A1}$ is an unbiased estimator of $\beta_{A(B)}$ but not the best unbiased estimator. The best unbiased estimator $\tilde{\beta}_{A(B)}$ of $\beta_{A(B)}$ assuming equal variances for Y_{ij} is

$$\tilde{\beta}_{A(B)} = (Y_{A4} + Y_{A7} - Y_{A1} - Y_{A5})/2.$$

Getting a similar expression to $\tilde{\beta}_{B(A)}$, we have

$$\tilde{\beta}_{A(B)} + \tilde{\beta}_{B(A)} = (Y_{A4} + Y_{A7} - Y_{A1} - Y_{A5} + Y_{B4} + Y_{B7} - Y_{B2} - Y_{B6}) / 2. \tag{6.8}$$

Clearly, the right-hand sides of Equations 6.5 and 6.8 are not the same, and by ignoring second-order availability effects, the best unbiased estimators of the sum of two availability effects $\beta_{A(B)}$ and $\beta_{B(A)}$ is not the same as the best unbiased estimator of the two-factor interaction α_{AB}. However, we have an unbiased estimator of $\beta_{A(B)} + \beta_{B(A)}$, the same as the best unbiased estimator of α_{AB}.

Since we are not assuming second-order availability effects, let us ignore the three-factor interaction in the ANOVA model.

The best unbiased estimator of α_{AB}, given by $\breve{\alpha}_{AB}$, can easily be verified to be

$$\alpha_{AB} = [4(Y_4 - Y_1 - Y_2) + 3(Y_3 + Y_7 - Y_5 - Y_6)] / 7$$

Clearly, $\alpha_{AB} \neq \tilde{\beta}_{A(B)} + \tilde{\beta}_{B(A)}$, and there is no relationship between first-order availability effects and two-factor interactions.

The DCEs also result in proportions. Using the convention used throughout this book, let L_{ij} be the log odds of the proportion choosing brand i in the jth choice set and p_{oj} be the proportion of people choosing the no-choice option in the jth set; i = A, B, C, $j = 1, 2, \ldots, 7$. In that case,

$$L_{ij} = \ln(p_{ij}/p_{oj}),\ i = \text{A, B, C};\ j = 1, 2, \ldots, 7,$$

$$L_j = \ln[(1 - p_{oj})/p_{oj}],\ j = 1, 2, \ldots, 7.$$

In this case, L_j is the log odds of selecting at least one of the brands of the choice set j. Clearly, with such responses, $L_j \neq \Sigma_i L_{ij}$.

With DCEs, the result

$$\hat{\alpha}_{ABC} = \hat{\beta}_{A(BC)} + \hat{\beta}_{B(AC)} + \hat{\beta}_{C(AB)}$$

$$\hat{\alpha}_{AB} = \hat{\beta}_{A(B)} + \hat{\beta}_{B(A)}$$

is not true.

6.2.2 Generating the Design Matrix for an Availability Design

The elements of the design matrix **D** are determined by the structure of the availability matrix **Av**. For example, the coding for $\mathbf{D_a(1)}$—the rows of the design matrix corresponding to first choice set of design matrix $\mathbf{Av\text{-}5_a}$ (Table 6.1, Panel D) are shown in Table 6.3, Panel A. Choices for alternative 1 are modeled as the own effect of the alternative plus the cross effects of alternatives 2, 3,

TABLE 6.3

Design Matrix for Availability Design $\mathbf{Av\text{-}5_a}$ (Table 6.1, Panel D)

		A		
1111–1	00000	00000	00000	00000
00000	1111–1	00000	00000	00000
00000	00000	1111–1	00000	00000
00000	00000	00000	1111–1	00000
		B		
00000	–11–111	00000	00000	00000
00000	00000	00000	–11–111	00000
00000	00000	00000	00000	–11–111

and 4 on brand 1 (Equation 6.3a). Alternative 5 is absent from the choice set. The bold numbers identify "own effect" parameters. The dependent variable for this row is the logit of the proportion picking alternative 1 in choice set 1 to the proportion picking the no-choice option in choice set 1. The second row has the coding for choices of the second alternative in the first choice set and so forth. There are four rows because the choice set consists of four alternatives.

The coding for $\mathbf{D_a(2)}$—the rows of the design matrix corresponding to the second row of $\mathbf{Av\text{-}5_a}$ (Table 6.1, Panel D)—is shown in Table 6.3, Panel B. The first row has the coding for the second alternative. The second row has the coding for the fourth alternative. The third row has the coding for the fifth alternative. The first and third alternatives are absent from the choice set.

The complete design matrix $\mathbf{D_a}$ is created by recursively appending the $\mathbf{D_a(k)}$'s corresponding to the rows of $\mathbf{Av\text{-}5_a}$. The matrix $\mathbf{D_a}$ corresponding to $\mathbf{Av\text{-}5_a}$ will have 16 such submatrices, the first two of which are provided in Table 6.3.

6.3 Portfolio Designs

Portfolio choice occurs when decision makers must assemble an ensemble of components from a domain of alternatives. The choice problem takes at least two forms. The first occurs when the decision maker must make choices between predetermined sets of components. This form is discussed by Louviere, Hencher, and Swait (2000) using a "city break" example from Dellaert, Borgers, and Timmermans (1995). The choice task is to assemble a get-away portfolio from the predetermined options of city, transportation mode, and hotel categories. One option must be selected from each preset category.

The second portfolio choice form occurs when "free format choices" are made of combinations from a set of predetermined alternatives. Decision makers are presented with choice sets t_i of alternatives, and they select $s_{ij} \in t_i$, where s_{ij} is the subset of choice set t_i selected by respondent j. For example, suppose an insurance company wished to create customized insurance packages that appeal to a variety of groups that differ in age, gender, marital status, presence of children, and so forth. Different combinations of coverage may appeal to different groups. Youths may prefer health club options, couples with children may prefer dental care, childless couples may want pregnancy coverage, middle-aged people may want long-term care coverage, and older people may want expanded drug coverage. However, constraints, such as cost, preclude offering every possibility. The question is, which subset of feasible offerings would maximize the attractiveness of the proposed insurance package? For this form of portfolio choice, the researcher would present respondents with subsets of component alternatives, and the experimental task is for the respondent to assemble a most preferred combination from each subset, including the possibility of selecting none of the alternatives in the set. The task may include constraints, such as picking "up to" 2 (i.e., 0, 1, or 2), up to 3, through up to the complete choice set.

Regardless of the task form, the composition of the chosen set of alternatives may affect the aggregate utility of the chosen combination. That is, the utility of each choice component is not invariant but in part depends on the presence or absence of the other chosen choice components. This is even more the case than with the pick-one availability problem because respondents may choose both of two complementary options.

6.3.1 Random Utility Model for Portfolio Choice

Consider a choice task for which the choice set consists of two components, A and B. The set of possible choices consists of A, B; AB, or none of the components, may be conceived as compound choices, $A \cap \bar{B}, \bar{A} \cap B, A \cap B, \bar{A} \cap \bar{B}$. Here, $\bar{A}$ is the complement of the set A. Assuming independence, choice probabilities for the various combinations of A and B may be calculated as the product of the probabilities of choosing (or not choosing) the components separately. For example, the probability of choosing none of the components is $P(\bar{A})* P(\bar{B})$, that is, the probability of not choosing A *and* not choosing B. The probability of choosing A equals $P(A)* P(\bar{B})$, where $P(A)$ is the probability of choosing A, and $P(\bar{B})$ is the probability of not choosing B (not-B), with similar interpretations for other alternatives in a choice set.

For illustration, suppose $v_A = 0.5$ and $v_B = -.40$. According to Equation 1.7 of Chapter 1, the probabilities of A versus NC are .62 (A) versus .38 (NC). The probabilities of B versus NC are .25 (B) and 0.75 (NC). The probabilities of the respective combinations are

$$P(A) = P(A) * P(\bar{B}) = .62 * .75 = .46$$

$$P(B) = P(\bar{A}) * P(B) = .38 * .25 = .10$$

$$P(AB) = P(A) * P(B) = .62 * .25 = .16$$

$$P(NC) = P(\bar{A}) * P(\bar{B}) = .38 * .75 = .28.$$

Note that the probability of choosing A from the marginal choice set consisting of A or $\bar{A}$ (.62) is equal to the sum of the joint probabilities for $P(A)$ and $P(AB)$, .46 + .16 = .62. The probability of choosing B from the choice set B or $\bar{B}$ (.25) is equal to the sum of the probabilities for $P(B)$ and $P(AB)$, .10 + .16 = .25 (subject to rounding error).

In the case of the choice of two alternatives, the four combinations may be represented as shown in Figure 6.1, which illustrates that the probability of the respective combinations of two alternatives corresponds to the quadrature of a bivariate distribution.

The example assumes that the cross effects for A and B are both positive, v_a and v_b, respectively, thus *increasing* the attractiveness of both alternatives when both are present in a choice set. An equivalent representation to increasing the value of A and B is to view their joint effect as shifting the origin of the distribution in the *opposite* direction, that is, $-v_a$ and $-v_b$ units: A *positive* cross effect results in a *negative* shift of the origin of the bivariate distribution, resulting in a larger proportion of the distribution above and to the right of the translated origin. It is then evident that increasing the

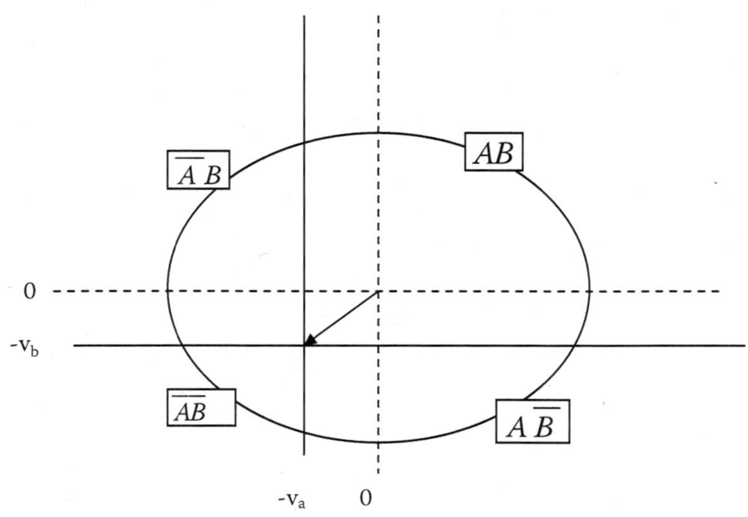

FIGURE 6.1
Combinations of choosing two alternatives from a set of four.

attractiveness of *both* A and B increases the probability of choosing both, and reduces the probability of choosing A or B singly, or NC. On the other hand, a positive value for A and a negative value for B (say v_a and $-v_b$) shifts the origin in a northwesterly direction, reducing the probability of *P*(AB) and *P*(B) and increasing *P*(A) and *P*(NC).

The amount of origin shift on a dimension occurring with the addition of B to a choice set may be viewed as a shift of the origin on the A axis, given by the B-on-A cross effect. A variety of observed choice proportions for the respective combinations may be accommodated with appropriate values of the cross effects. Generalizing, choice probabilities for the alternatives are modeled by origin shift in a three-dimensional distribution, four alternatives by origin shift in a four-dimensional distribution, and so forth.

Since cross effects may be estimated using availability designs, with appropriate coding, the designs based on Hadamard matrices may be used to estimate portfolio choice for brand-only models. Consider, for example, the design shown in Table 6.4. The first choice set contains only alternative D and NC. Choice set 2 contains the alternatives {A, B, D}. Given the options of A, B, and D, respondents may choose A, B, D, AB, AD, BD, ABD, and no choice.

Table 6.5 illustrates coding for portfolio choice data using the design in Table 6.4. There are four blocks of columns, labeled A, B, C, and D. Beneath each block are four columns labeled A, B, C, and D and containing the coding for the own and cross effects. For example, the column labeled A under the A block contains coding for the own effect of A. The coding under the B column contains the coding for the B cross effect on A, and so on. The coding under the A column of the B block of coding contains coding for the B-on-A cross effect, the coding under the B column contains coding for the own effect of B, and so on.

TABLE 6.4

Availability Design for Portfolio Choice

Choice		Design				
Set	Composition	A	B	C	D	NC
1	D	–1	–1	–1	1	1
2	ABD	1	1	–1	1	1
3	ACD	1	–1	1	1	1
4	BCD	–1	1	1	1	1
5	ABC	1	1	1	–1	1
6	B	–1	1	–1	–1	1
7	C	–1	–1	1	–1	1
8	A	1	–1	–1	–1	1

TABLE 6.5
Coding for Portfolio Design (Table 6.4)

		Own and Cross Effects																		
		0.5	0.2	−0.2	0.1	−0.1	0.4	0.1	0.05	0.05	−0.05	0.3	0.1	−0.1	−0.2	0.25	0.2			
		A				B				C				D						
Set	Comb.	A	B	C	D	A	B	C	D	A	B	C	D	A	B	C	D	Logit	P(comb)	P(nc)
D	D	0	0	0	0	0	0	0	0	0	0	0	0	0	0	0	1	0.20	0.55	0.45
ABD	A	1	−1	0	−1	0	0	0	0	0	0	0	0	0	0	0	0	0.20	0.08	
	B	0	0	0	0	−1	1	0	−1	0	0	0	0	0	0	0	0	0.45	0.11	
	D	0	0	0	0	0	0	0	0	0	0	0	0	−1	−1	0	1	0.50	0.11	
	AB	1	1	0	−1	1	1	0	−1	0	0	0	0	0	0	0	0	0.85	0.16	
	AD	1	−1	0	1	0	0	0	0	0	0	0	0	1	−1	0	1	0.70	0.14	
	BD	0	0	0	0	−1	1	0	1	0	0	0	0	−1	1	0	1	0.65	0.13	
	ABD	1	1	0	1	1	1	0	1	0	0	0	0	1	1	0	1	1.05	0.20	0.07
ACD	A	1	0	−1	−1	0	0	0	0	0	0	0	0	0	0	0	0	0.20	0.09	
	C	0	0	0	0	0	0	0	0	−1	0	1	−1	0	0	0	0	0.15	0.08	
	D	0	0	0	0	0	0	0	0	−1	0	−1	1	−1	0	−1	1	0.25	0.09	
	AC	1	0	1	−1	0	0	0	0	1	0	1	−1	0	0	0	0	0.85	0.17	
	AD	1	0	−1	1	0	0	0	0	0	0	0	0	1	0	−1	1	0.70	0.15	
	CD	0	0	0	0	0	0	0	0	−1	0	1	1	−1	0	1	1	0.45	0.11	
	ACD	1	0	1	1	0	0	0	0	1	0	1	1	1	0	1	1	1.15	0.23	0.07

BCD	B	0	0	0	0	0	1	−1	−1	0	0	0	0	0	0	0	0	0.25	0.09	
	C	0	0	0	0	0	0	0	0	0	−1	1	−1	0	0	0	0	0.25	0.09	
	D	0	0	0	0	0	0	0	0	0	0	0	0	0	−1	−1	1	0.15	0.08	
	BC	0	0	0	0	0	1	1	−1	0	1	1	−1	0	0	0	0	0.60	0.13	
	BD	0	0	0	0	0	1	−1	1	0	0	0	0	0	1	−1	1	0.10	0.08	
	CD	0	0	0	0	0	0	0	0	0	−1	1	1	0	−1	1	1	1.10	0.22	
	BCD	0	0	0	0	0	1	1	1	0	1	1	1	0	1	1	1	1.15	0.23	0.07
ABC	A	1	−1	−1	0	0	0	0	0	0	0	0	0	0	0	0	0	0.50	0.11	
	B	0	0	0	0	−1	1	−1	0	0	0	0	0	0	0	0	0	0.40	0.10	
	C	0	0	0	0	0	0	0	0	−1	−1	1	0	0	0	0	0	0.30	0.09	
	AB	1	1	−1	0	1	1	−1	0	0	0	0	0	0	0	0	0	1.10	0.19	
	AC	1	−1	1	0	0	0	0	0	1	−1	1	0	0	0	0	0	0.50	0.11	
	BC	0	0	0	0	−1	1	1	0	−1	1	1	0	0	0	0	0	0.80	0.14	
	ABC	1	1	1	0	1	1	1	0	1	1	1	0	0	0	0	0	1.20	0.21	0.06
B	B	0	0	0	0	0	1	0	0	0	0	0	0	0	0	0	0	0.40	0.60	0.40
C	C	0	0	0	0	0	0	0	0	0	0	1	0	0	0	0	0	0.30	0.57	0.43
A	A	1	0	0	0	0	0	0	0	0	0	0	0	0	0	0	0	0.50	0.62	0.38

There are eight blocks of row coding in the table corresponding to the choice sets shown in Table 6.4. The first row labeled D contains coding for the choice set {D, NC}. The second block of rows contains coding for the choice set {A, B, D, NC}. Note interpretation of coding for the portfolio choice problems differs from the pick-one availability problem. With the availability problem, the cross effect is interpreted as the effect of the availability of an alternative in the choice set even though it is not picked. In the portfolio problem, the coding is interpreted as the coding for the value of an option provided it is part of a selected portfolio. For example, coding for the first row of choice set 2 is for choosing A (1), $\bar{B}$ (−1), and $\bar{D}$ (−1). Option C is not present in the set and is coded 0.

The fourth row contains coding for the proportion picking the AB combination from the set {A, B, D}. People selecting this combination receive the value of A and B. The value of A in this set is given by the own effect of A (1) adjusted by cross effects of B on A (1) and $\bar{D}$ on A (−1). The value of B in this set is given by the own effect of B (1) adjusted by the cross effect of A on B (1) and the cross effect of $\bar{D}$ on A (−1). Similar coding applies to the remaining rows.

The dependent variable in the analysis is the logit of the proportion choosing each permitted combination in the choice set divided by the proportion choosing no choice. A set of own and cross effects is provided at the top of Table 6.5. For example, the own effect for A is .50, and the cross effects are AB (.2), AC (−.2), and AD (.1). The own effect for B is (.4), and the cross effects are BA (−.1), BC (.1), BD (.05), and so on. Given these values, the logit value for picking A from the set {A, B, D} is equal to .20, which equals ln(.08/.07). The net shift of origin on the A (horizontal) dimension of Figure 6.1 is the sum of the cross effects of other alternatives in the set on A. The net shift of origin on the B (vertical) dimension of Figure 6.1 is the sum of the cross effects of alternatives in the set on B. For designs with more alternatives, the effect of other alternatives in a choice set is interpreted as the sum of the cross effects in a space of higher dimension.

In general, Hadamard matrices may be used to formulate portfolio choice problems corresponding to the pick-one problems discussed in Section 6.2. When there are numerous options, however, the number of combinations in choice sets with many options will be quite large, and it is likely that some of these combinations will receive no choices, resulting in zero choice probabilities. In these circumstances, it is desirable, if not necessary, to restrict choice to smaller combinations, such as pick up to two or pick up to three alternatives. "Pick at least two" is the minimum that allows complementary behavior to be observed. "Pick at least three" allows lack of fit for the model to be evaluated. That is, the effects may be estimated using only the pick-at-least-two data, and the observed proportions for combinations of three used to evaluate fit of the model.

6.4 Brand and One (or More) Attributes

Consider now when there is one attribute with two levels associated with each of the m brands. The cross-effects model equivalent to Equation 6.1 having availability effects and effects for a single attribute with two levels may be represented as

$$E(V_{i=1|k=1} - V_{nc}) = \alpha_{11} + \alpha_{21} + \alpha_{31} + \alpha_{41} + \alpha_{51} + \alpha_{61} + \pi_{11} + \pi_{21} + \pi_{31} + \pi_{41} + \pi_{51} + \pi_{61}, \tag{6.9a}$$

$$E(V_{i=1|k=2} - V_{nc}) = \alpha_{11} + \alpha_{21} - \alpha_{31} + \alpha_{41} + \alpha_{51} + \alpha_{61} + \pi_{11} + \pi_{21} - \pi_{31} + \pi_{41} + \pi_{51} + \pi_{61}. \tag{6.9b}$$

Own and cross effects of a four brand with a one attribute at two levels can be organized as follows:

$$V_1 = \begin{bmatrix} \alpha_{11} & \alpha_{12} & \alpha_{13} & \alpha_{14} \\ \alpha_{21} & \alpha_{22} & \alpha_{23} & \alpha_{24} \\ \alpha_{31} & \alpha_{32} & \alpha_{33} & \alpha_{34} \\ \alpha_{41} & \alpha_{42} & \alpha_{43} & \alpha_{44} \end{bmatrix}, \quad V_2 = \begin{bmatrix} \pi_{11} & \pi_{12} & \pi_{13} & \pi_{14} \\ \pi_{21} & \pi_{22} & \pi_{23} & \pi_{24} \\ \pi_{31} & \pi_{32} & \pi_{33} & \pi_{34} \\ \pi_{41} & \pi_{42} & \pi_{43} & \pi_{44} \end{bmatrix}.$$

The elements of V_1 have the same interpretation as in the brand-only form. The parameters of V_2 along the main diagonal are the own effects of the *attribute*. For example, if the attribute is price, elements π_{ii} along the main diagonal are the effects of the own price of the alternative on the log odds of its being chosen. The elements along row i of V_2 are the effects of the prices of the other alternatives on the log odds of choosing the brand. There are 2^*m^2 parameters. Research on designs from which attribute cross effects may be estimated only is at the beginning stages. The following sections of this chapter discuss some currently available approaches.

From a design standpoint, the problem of attaching one or more attributes to an availability design is that the attributes may assume values *only* in the case that the brand is present in the choice set. It is not feasible, for example, to treat the problem as a factorial design in which one factor is "presence/absence brand A" and another factor is "price brand A." Inevitably, the second factor would be giving a price for a brand when the first factor indicated it was absent.

The parameters that must be estimated increase rapidly as more than one attribute are included in a design. For example, a brand-only design has m^2 parameters, a one-attribute design at two levels has 2^*m^2 parameters, a two-attribute design at two levels has 3^*m^2 attributes, and so forth. Parameters increase even more rapidly when there are more than two levels of each attribute. With one attribute at three levels, there are 3^*m^2 parameters; with two attributes at three levels there are 5^*m^2 parameters. One of the main tasks when there are two or more attributes is devising strategies for reducing the number of parameters to be estimated.

6.4.1 Brands with One Attribute

Two articles have considered the problem of attaching an attribute at *two levels* to an availability design. Raghavarao and Wiley (1994) present two approaches that might be called the "small choice set size" and the "large choice set size" strategies. The small choice set size approach entails taking $2m$ choice sets of size 1, $m(m - 1)$ choice sets of size 2, and $m(m - 1)(m - 2)/2$ choice sets of size 3. This approach requires a large number of choice sets. The total number of sets (and parameters to be estimated) for $m = 3, 4, 5$, and 6 brands is 15 (18), 32 (32), 60 (50), and 102 (72), respectively. However, in some situations, such as when the choice experiment will be computer (or Internet) based, with alternatives presented on a computer screen, small choice set designs may be the only ones feasible. If large choice sets are feasible, large choice set designs with only $2m + 1$ choice sets may be constructed by creating sets with sizes $m - 1$ and m. The number of choice sets for 3, 4, 5, and 6 brands are 7, 9, 11, and 13, respectively.

Middle-range choice set sizes may be constructed using mutually orthogonal Latin squares (MOLS). For example, designs for $m \leq 6$ brands with one attribute at four levels having 25 choice sets may be constructed by selecting any m rows from the MOLS of order 5 provided in Table 6.6.

For example, the design for three brands with one attribute at four levels shown in Table 6.7 is constructed using the transpose of the first three rows of Table 6.6. Zeros signify the brand is missing from the choice set. Binary coding for levels 1 through 4 is $1 \rightarrow \{1\ 0\ 0\}$, $2 \rightarrow \{0\ 1\ 0\}$, $3 \rightarrow \{0\ 0\ 1\}$, $4 \rightarrow \{-1\ -1\ -1\}$. Asterisks indicate that the brand is missing from the set.

Lazari and Anderson (1994) provide a related approach. Designs for m brands at $s - 1$ levels are created based on s^m fractional factorial designs. As with the previous approach, the absence of a brand is represented as level 0 of the attribute, and levels 1 through s signify presence at levels 1, 2, … , s, respectively. Brands will be absent in $1/s$ of the choice sets and present at each level in $1/s$ sets.

Lazari and Anderson (1994) provide a catalog of designs for designs with 4, 5, 7, 8, and 9 levels and a variety of numbers of brands and choice sets. Efficiencies for the designs are also provided. An empirical application of the approach illustrating a design having 12 brands at two price levels is provided.

With the Lazari and Anderson (1994) approach, it may be desirable to add a second absent condition and use 4^m factorial designs to get better availability balance. Orthogonal fractional plans for 3, 4, 5, and 6 brands based on

TABLE 6.6

Array with 25 Columns and 6 Rows

	Choice Set																								
Alternative	**1**	**2**	**3**	**4**	**5**	**6**	**7**	**8**	**9**	**10**	**11**	**12**	**13**	**14**	**15**	**16**	**17**	**18**	**19**	**20**	**21**	**22**	**23**	**24**	**25**
A	0	1	2	3	4	0	1	2	3	4	0	1	2	3	4	0	1	2	3	4	0	1	2	3	4
B	0	2	4	3	1	1	3	0	4	2	2	4	1	0	3	4	1	3	2	0	3	0	2	1	4
C	0	2	4	3	1	3	0	2	1	4	1	3	0	4	2	2	4	1	0	3	4	1	3	2	0
D	0	2	4	3	1	4	1	3	2	0	3	0	2	1	4	1	3	0	4	2	2	4	1	0	3
E	0	2	4	3	1	2	4	1	0	3	4	1	3	2	0	3	0	2	1	4	1	3	0	4	2
F	0	0	0	0	0	1	1	1	1	1	2	2	2	2	2	3	3	3	3	3	4	4	4	4	4

TABLE 6.7

Design for Three Brands with One Attribute at Four Levels

Choice Set	L (A)	L(B)	L(C)	Alternatives			Binary Coding (Choice)									Orthogonal Polynomial Coding (Choice)								
							A			B			C			A			B			C		
				A	B	C										L	Q	C	L	Q	C	L	Q	C
1	*	*	*	0	0	0	*	*	*	*	*	*	*	*	*	*	*	*	*	*	*	*	*	*
2	LA2	LB2	LC2	1	2	2	1	0	0	0	1	0	0	1	0	−3	1	−1	−1	−1	3	−1	−1	3
3	LA3	LB3	LC3	2	4	4	0	1	0	−1	−1	−1	−1	−1	−1	−1	−1	3	3	1	1	3	1	1
4	LA4	LB4	LC4	3	3	3	0	0	1	0	0	1	0	0	1	1	−1	−3	1	−1	−3	1	−1	−3
5	LA5	LB5	LC5	4	1	1	−1	−1	−1	1	0	0	1	0	0	3	1	1	−3	1	−1	−3	1	−1
6	*	LB6	LC6	0	1	3	*	*	*	1	0	0	0	0	1	*	*	*	−3	1	−1	1	−1	−3
7	LA7	LB7	*	1	3	0	1	0	0	0	0	1	*	*	*	−3	1	−1	1	−1	−3	*	*	*
8	LA8	*	LC8	2	0	2	0	1	0	*	*	*	0	1	0	−1	−1	3	*	*	*	−1	−1	3
9	LA9	LB9	LC9	3	4	1	0	0	1	−1	−1	−1	1	0	0	1	−1	−3	3	1	1	−3	1	−1
10	LA10	LB10	LC10	4	2	4	−1	−1	−1	0	1	0	−1	−1	−1	3	1	1	−1	−1	3	3	1	1
11	*	LB11	LC11	0	2	1	*	*	*	0	1	0	1	0	0	*	*	*	−1	−1	3	−3	1	−1
12	LA12	LB12	LC12	1	4	3	1	0	0	−1	−1	−1	0	0	1	−3	1	−1	3	1	1	1	−1	−3
13	LA13	LB13	*	2	1	0	0	1	0	1	0	0	*	*	*	−1	−1	3	−3	1	−1	*	*	*
14	LA14	*	LC14	3	0	4	0	0	1	*	*	*	−1	−1	−1	1	−1	−3	*	*	*	3	1	1
15	LA15	LB15	LC15	4	3	2	−1	−1	−1	0	0	1	0	1	0	3	1	1	1	−1	−3	−1	−1	3
16	*	LB16	LC16	0	4	2	*	*	*	−1	−1	−1	0	1	0	*	*	*	3	1	1	−1	−1	3
17	LA17	LB17	LC17	1	1	4	1	0	0	1	0	0	−1	−1	−1	−3	1	−1	−3	1	−1	3	1	1
18	LA18	LB18	LC18	2	3	1	0	1	0	0	0	1	1	0	0	−1	−1	3	1	−1	−3	−3	1	−1
19	LA19	LB19	LC19	3	2	0	0	0	1	0	1	0	*	*	*	1	−1	−3	−1	−1	3	*	*	*
20	LA20	*	LC20	4	0	3	−1	−1	−1	*	*	*	0	0	1	3	1	1	*	*	*	1	−1	−3

21	LA21	LB21	LC21	0	3	4	*	*	*	0	0	1	−1	−1	−1	*	*	*	1	−1	−3	3	1	1
22	LA22	LB22	*	1	0	1	1	0	0	*	*	*	1	0	0	−3	1	−1	*	*	*	−3	1	−1
23	LA23	LB23	LC23	2	2	3	0	1	0	0	1	0	0	0	1	−1	−1	3	−1	−1	3	1	−1	−3
24	LA24	LB24	LC24	3	1	2	0	0	1	1	0	0	0	1	0	1	−1	−3	−3	1	−1	−1	−1	3
25	LA25	LB25	*	4	4	0	−1	−1	−1	−1	−1	−1	*	*	*	3	1	1	3	1	1	*	*	*

*The brand is missing from the set.

4^m factorial designs are available with the number of choice sets $K = 16$, 16, 16, 16, and 25, respectively. Brands occur in one-half of the K choice sets in a design. Wiley (1997) provides an example of how data from a one attribute at a two-level study may be used by a decision support system to support a pricing decision.

6.4.2 One Attribute at Three or More Levels

Both Raghavarao and Wiley (1994) and Lazari and Anderson (1994) provide approaches for estimating availability models with one attribute at three or more levels. Raghavarao and Wiley restrict attention to three levels and show that the $2m^2 + m$ parameters may be estimated with designs that have $3m$ choice sets of size 1, $2m(m - 1)$ choice sets of size 2, and $m(m - 1)(m - 2)$ choice sets of size 3. The number of choice sets (and numbers of parameters) for $m = 3$, 4, 5, and 6 with this approach are 27 (21), 60 (36), 115 (55), 198 (78), and 315 (105), respectively.

Lazari and Anderson (1994) propose using s^m fractional factorial designs, where $s = 0$ is the absent condition and $s = 1, \ldots, s - 1$ are the levels of the attribute to be included in the design. For a three-level design, set sizes for 3, 4, 5, and 6 brands based on 4^m orthogonal fractional designs are available with $K = 16$, 16, 16, 16, and 25, respectively. However, to be estimable $3m \leq 2/3K$, and this condition is met only for $m = 3$. Again, to get better balance in the design, additional absent conditions may be added. Lazari and Anderson provide designs for the following conditions: $K = 25$, $s = 5$, $m \leq 5$; $K = 32$, $s = 4$, $m \leq 8$; $K = 50$, $s = 5$, $m \leq 10$; and $K = 64$, $s = 8$, $m \leq\leq 8$;

The number of parameters becomes very large as levels are added to an attribute. There are m availability and $m(s - 1)$ own and cross-attribute parameters for each of m brands, or m^2s parameters total. To provide models with interpretable parameters, both Raghavarao and Wiley (1994) and Lazari and Anderson (1994) estimate orthogonal polynomials when $s \geq 3$. Both note that polynomials of order greater than 3 are difficult to interpret and restrict attention to linear, quadratic, and cubic effects. The model that can be used is:

$$v_{i\in t} = \alpha_i + \beta_{1i}Lin_i + \beta_{2i}Quad_i + \beta_{3i}Cub_i + \cdots$$
$$\Sigma_{i'\neq i}([\gamma_{1ii'}Lin_{i'} + \gamma_{2ii'}Quad_{i'} + \cdots] + \delta_{ii'}Z_{i'})$$

where

α_i = intercept for brand i,
β_{1i} = attribute linear effect for brand i,
β_{2i} = attribute quadratic effect for brand i,
β_{3i} = attribute cubic effect for brand i,

⋮ ⋮

$\gamma_{1ii'}$ = attribute linear cross effect of brand i' on brand i,
$\gamma_{2ii'}$ = attribute quadratic cross effect of brand i' on brand i,

$\gamma_{3ii'}$ = attribute cubic cross effect of brand i' on brand i,

$\vdots$

$\delta_{ii'}$ = availability cross effect of brand i' on brand i.

For $s = 3$, Raghavarao and Wiley model *own price effects* with linear and quadratic terms (an own, cubic effect term would be added for $s > 3$). Only linear terms are estimated for price cross effects. The sign of the quadratic own effect indicates whether preference decreases (presumably) with increasing price at an increasing or decreasing rate. Cubic terms generally get at symmetry around an "inflection point." The signs of linear cross-price effects indicate whether the magnitude of the own, linear, quadratic, and cubic effects increase or decrease as the competing brand increases its own price. With models in this form, there are m $(2m + 3)$ parameters to be estimated. For example, with $m = 5$ and $s = 5$, there would be 225 parameters in a complete model and 65 in the Raghavarao and Wiley restricted model.

Lazari and Anderson (1994) take a somewhat different approach to model formulation. They include linear through cubic, own, and cross effects as *potential* components of a model. They then estimate only a "relevant" subset of potential cross effects. For example, they provide an example in which a new frozen food line is to be introduced into a market consisting of five competing lines, including another line of the firm. Each existing line comes in beef and chicken *forms*, as does the new line. The *brand line* and the *form* attributes were represented as 12 brand line/form pseudobrands as discussed, and a single, two-level price attribute was evaluated. Even with $s = 2$, a complete design would require 288 parameters. The objective of the study, however, is to estimate demand for the new line and determine the impact its introduction would have on competitors, including the other line of the firm (i.e., cannibalization). Given this objective, Lazari and Anderson reason that only the cross effects of the new brand on other brands need to be estimated (all own effects were estimated). With this restriction, the model of interest has 128 parameters.

6.5 Brands and More than One Attribute

Little work has been done on availability designs with two or more attributes. Anderson et al. (1992) suggest a composite of several designs. They created an availability design by first creating a fractional factorial for the condition for which all brands are present. They then appended designs for the brands taken two at a time. Using the six-brand example discussed here as an example, a design with three attributes at two levels would start with a fractional design for the 2^{18} all-brands-present condition. An orthogonal design in 20 runs is available. Next, fractional designs for each of the 15 pairs of brands are appended. An orthogonal fraction for the 2^6 pairs design in eight runs is available. The complete design consists of 20 + 8*15 = 140 choice sets. Each brand will be present in each of first 20 choice sets and $(m - 1)*8 = 40$ of the 15 choices sets that pair the brands, where m = the number of brands. Each brand is present in 60 (20 + 40) choice sets and absent in 80 choice sets. It is unlikely that a single respondent would be given all choice sets. A variety of sampling strategies may be used to select a predetermined subset of profiles for presentation to individuals, such as simple random sampling of profiles from the complete design or stratified sampling using the respective fractional designs as strata. Alternatively, a blocking factor could be introduced and respondents randomly assigned to blocks. Each respondent may then be randomly assigned to one of the blocks of 35 profiles.

The number of choice sets increases rapidly with additional levels using this approach. A design with three attributes at three levels would start with a fractional design for the 3^{18} all-brands-present condition. A design in 64 runs is available. Fractions of the 3^6 designs for the 15 pairs are available in 18 runs. The complete design consists of 64 + 18*15 = 334 choice sets.

6.5.1 General Results for Two or More Attributes with Two or More Levels

Let there exist t mutually orthogonal Latin squares of order $s \times s$, in symbols 0, 1, ... , $s - 1$. Add two more squares to that set, the first with i in the ith row of the first and j in the jth column in the second square, $i = 0, 1, \ldots, s - 1; j = 0, 1, \ldots, s - 1$. Superimpose all the $t + 2$ squares and write an $s^2 \times (t + 2)$ array A corresponding to the entries in each of the cells. Suppose we have m brands ($m \le (t + 2)$) and two attributes at p and q levels, where $(s - 1) = pq$. Choose m columns of the $t + 2$ columns of A and identify the selected columns to the brands. In the selected columns, 0 denotes absence of the brand and 1, 2, ... , $s - 1$ are identified with (u, w) levels of the two attributes, $u = 0, 1, 2, \ldots, p - 1$ and $w = 0, 1, 2, \ldots, q - 1$, to get m brands DCE with two attributes at p and q levels, respectively.

Example 6.1

Designs for up to six alternatives with two attributes at two levels in 24 choice sets may be constructed using MOLS of order 5 coded as 0, 1, 2, 3, 4 (Table 6.6). The transpose of the display in Table 6.6 is the **A** matrix mentioned in the paragraph preceding this example. For example, the transpose of any subset of rows $m \le 6$ of Table 6.6 may be selected. Attributes are coded as follows with zero signifying that the alternative is missing:

		Attribute	
Code		A	B
0	→	Missing	
1	→	0	0
2	→	0	1
3	→	1	0
4	→	1	1

In Table 6.8, we provide the choice sets.

TABLE 6.8

Matrix **A** and Choice Sets with Six Brands and Two Attributes

Design Matrix						Choice Set				
0	0	0	0	0	0					
1	2	2	2	0	0	$\{a_{00}$	b_{01}	c_{01}	$d_{01}\}$	
2	4	4	4	4	0	$\{a_{01}$	b_{11}	c_{11}	d_{11}	$e_{11}\}$
3	3	3	3	3	0	$\{a_{10}$	b_{10}	c_{10}	d_{10}	$e_{01}\}$
4	1	1	1	1	0	$\{a_{11}$	b_{00}	c_{00}	d_{00}	$e_{00}\}$
0	1	3	4	2	1	$\{b_{00}$	c_{10}	d_{11}	e_{01}	$f_{00}\}$
			...							
3	1	2	0	4	4	$\{a_{01}$	b_{00}	c_{01}	e_{11}	$f_{11}\}$
4	4	0	3	2	4	$\{a_{11}$	b_{11}	d_{10}	e_{01}	$f_{11}\}$

TABLE 6.9

Three Superimposed Squares of Order 7

000,	101,	202,	303,	404,	505,	606,
110,	211,	312,	413,	514,	615,	016,
220,	321,	422,	523,	624,	025,	126,
330,	431,	532,	633,	034,	135,	236,
440,	541,	642,	043,	144,	245,	346,
550,	651,	052,	153,	254,	355,	456,
660,	061,	162,	263,	364,	465,	566,

TABLE 6.10

Matrix **A** and Choice Sets

Design Matrix A			Choice Sets
0	0	0	Empty set
1	0	1	$\{a_{00}, c_{00}\}$
2	0	2	$\{a_{01}, c_{01}\}$
3	0	3	$\{a_{02}, c_{02}\}$
4	0	4	$\{a_{10}, c_{10}\}$
5	0	5	$\{a_{11}, c_{11}\}$
6	0	6	$\{a_{12}, c_{12}\}$
...	...	...	...
6	6	0	$\{a_{12}, b_{12}\}$
0	6	1	$\{b_{12}, c_{00}\}$
1	6	2	$\{a_{00}, b_{12}, c_{01}\}$
2	6	3	$\{a_{01}, b_{12}, c_{02}\}$
3	6	4	$\{a_{02}, b_{12}, c_{10}\}$
4	6	5	$\{a_{10}, b_{12}, c_{11}\}$
5	6	6	$\{a_{11}, b_{12}, c_{12}\}$

Example 6.2

Designs with three brands and two attributes at two and three levels in 48 choice sets may be constructed using the three superimposed squares of order 7 shown in Table 6.9. Table 6.10 illustrates the coding of the design matrix **A** using the following coding:

		Attribute	
Code		**A**	**B**
0	→	Missing	
1	→	0	0
2	→	0	1
3	→	0	2
4	→	1	0
5	→	1	1
6	→	1	2

7

Sequential Methods

7.1 Introduction

Sequential methods of experimentation involve conducting a study in different stages using the information available in previous stages in planning and executing the further experimentation. Sequential methods are particularly timely because, with the advent of Web-based surveying, choice experiments will increasingly be computer assisted. An advantage of computer-assisted discrete choice experimentation (DCE) is that it may be done in real time; that is, prior choices may be used to compose subsequent choice sets. Consequently, it will be possible to modify subsequent choice set composition based on earlier choices. In this chapter, we provide design strategies for sequentially augmenting choice experiments to diagnose whether two-way or three-way interactions are needed and for determining which components should be included in the choice model.

Strategies for using prior information may be considered to lie on a continuum. At one extreme, information gained from a single choice from a prior choice set may be used to select the next choice set. At the other extreme, information gained from a prior *block* of choices may be used to select the next block of choice sets. The strategy we discuss in this chapter, except Section 7.4, follows the latter strategy; information gained from a prior block of choices is used to determine whether to go on to the next block of choice sets. A main-effects plan is used at the beginning, and if the model fits the data, the experimentation is stopped, and inferences are drawn on the main effects. If the main-effects model indicates lack of fit, further experimentation is made and a model of main effects and two factor interactions is used on the combined data and if there is no lack of fit, the study is terminated and inferences are drawn on the main effects and two-factor interactions. If the model of main effects and two-factor interactions indicates lack of fit, further data are collected to estimate three-factor interactions, and at this stage the study is terminated and inferences are drawn on main effects and two- and, three-factor interaction of the attributes. These results are discussed in Section 7.2 and are useful in every symmetric s^m experiment. We also discuss the use of a screening experiment to identify the factors to be included in the larger study.

In Section 7.3, we focus on symmetric 2^m experiments. To estimate all main effects and two- and three-attribute interactions using procedures discussed in Section 7.2, we need many profiles to collect the data. Usually, all two- and three-attribute interactions are not of interest to the researcher. However, price often is used as an attribute in marketing and economics research; in that case, the two-attribute interaction with price as one attribute is of interest. Let us also add a three-attribute interaction with price as one attribute. We may plan a study to estimate all main effects and if the main-effects plan shows lack of fit, we add more profiles to estimate two-attribute interactions with a common attribute and a three-attribute interaction involving the same attribute. We study these experiments in Section 7.3.

In Section 7.4, we consider a case for which information about a respondent's choice model is built up profile by profile. Although this monograph is mainly devoted to DCE, we consider a conjoint 2^m analysis (CA) study with attributes $A_1, A_2, \ldots, A_m$. We can sequentially present the profiles to the panel of respondents so that we can estimate the main effects of the attributes $A_1, A_2, \ldots, A_m$ one by one and a two-attribute interaction, say, A_1A_2.

In an s^m experiment, sometimes none of the main effects of the m attributes may be significant. In that case, the researcher may wish to abandon the study as early as possible. This setting is similar to interim analysis in clinical trials. If we use two choice sets with N respondents evaluating each of the choice sets, we may perform interim analysis with $N/2$ respondents evaluating each choice set, assuming N to be even. With interim analysis, we test whether the collected data indicate any evidence against the null hypothesis that planned attribute effects are zero. If there is evidence that no effects are significant, we terminate the study; otherwise, we complete the study to reach the conclusion. These results are presented in Section 7.5.

Sequential experiments for symmetric 3^m experiments to estimate main effects and a two-factor interaction are discussed in Section 7.6.

7.2 Sequential Experiment to Estimate All Two- and Three-Attribute Interactions

The organization of choice sets into blocks, combined with sequential testing, allows a specification search for model structure, that is, deciding whether interactions should be included in a model. Computer-administered research, however, implies limited screen "real estate," which limits the number of alternatives that may be presented in a single choice set. In addition, it is easy for respondents to "break off" Internet-based surveys. Consequently, if the potential of computer-assisted sequential choice experimentation is to be realized, we need the following capabilities:

1. We need a way to identify subsets in which main effects; main-effects and two-way interactions; and main-effects, two-way, and (potentially) higher-order interactions may be *sequentially* estimated.
2. Given such subsets, we need a testing strategy to decide whether it is necessary to go on to the next stage, that is, to add subsets to estimate higher-order interactions. A test of overall model fit is central to developing such a strategy.
3. When a design includes attributes that are costs or benefits (as often is the case), it is desirable that choice sets not include dominated or dominating alternatives. We need a way to generate Pareto optimal (PO) choice sets, that is, choice sets that do not include dominated or dominating alternatives. Procedures for creating PO subsets are discussed in Chapter 4.
4. If the subsets in items 1–3 are too large to give in a single presentation (e.g., because they will be presented on a computer screen), we need a strategy for breaking them up into smaller subsets. Procedures for breaking large choice sets into smaller subsets are discussed in Chapter 5.

This section discusses procedures that provide the first two of these capabilities.

Raghavarao and Wiley (2006) prove the following theorems for a symmetrical s^m experiment in which s is an integer at least 2.

Theorem 7.1

(a) We can estimate all m main $= m(s - 1)$ contrasts of main effects by using two consecutive choice sets S_ℓ and $S_{\ell+1}$, where $(s - 2) \le \ell \le (m - 1)(s - 1)$.

(b) We can estimate all m two-way $= m(m - 1)(s - 1)^2/2$ contrasts of two-factor interactions with three consecutive choice sets S_ℓ, $S_{\ell+1}$, $S_{\ell+2}$ where $2(s - 2) \le \ell \le (m - 2)(s - 1)$.

(c) We can estimate all m three-way $= m(m - 1)(m - 2)(s - 1)^3/6$ contrasts of three-factor interactions with four consecutive choice sets S_ℓ, $S_{\ell+1}$, $S_{\ell+2}$, $S_{\ell+3}$ where $3(s - 2) \le \ell \le (m - 3)(s - 1)$.

In Theorem 7.1, a choice set S_ℓ consists of profiles $x_1, x_2, \ldots, x_m$ where $\sum x_i = \ell$. As subsets of PO choice sets are PO, if the choice sets discussed in Theorem 7.1 have too many profiles for respondents to evaluate, they may be subdivided into choice sets with fewer profiles so that respondents may make a suitable evaluation. All the choice sets contain a no-choice option.

In a 3^5 experiment, choice sets S_ℓ and $S_{\ell+1}$, where $1 \le \ell \le 8$ are needed to estimate main effects; choice sets S_ℓ, $S_{\ell+1}$, $S_{\ell+2}$ where $2 \le \ell \le 6$ are needed to estimate main effects and all two-factor interactions; choice sets S_ℓ, $S_{\ell+1}$, $S_{\ell+2}$, $S_{\ell+3}$ where $3 \le \ell \le 4$ are needed to estimate main effects, two-factor, and three-factor interactions. While planning a study that may continue until three-factor interactions are estimated, we need to choose ℓ satisfying $1 \le \ell \le 8$; $2 \le \ell \le 6$; and $3 \le \ell \le 4$. Hence, we choose $\ell = 3$ or $\ell = 4$; let us take $\ell = 3$. The choice sets

$$S_3 = \{(x_1, x_2, x_3, x_4, x_5) | x_i = 1, x_j = 1, x_k = 1; i \ne j \ne k; i, j, k = 1, 2 \ldots, 5\}$$

$$\cup [(x_1, x_2, x_3, x_4, x_5)] | x_i = 1, x_j = 2, i \ne j; i, j = 1, 2, \ldots, 5$$

$$S_4 = \left\{(x_1, x_2, x_3, x_4, x_5) \middle| \begin{matrix} x_i = 1, x_j = 1, x_k = 1, x_m = 1; \\ i \ne j \ne k \ne m; i, j, k, m = 1, 2, \ldots, 5 \end{matrix} \right\}$$

$$\cup \{(x_1, x_2, x_3, x_4, x_5) | x_i = 2, x_j = 1, x_k = 1; i \ne j \ne k; i, j, k = 1, 2, \ldots, 5\}$$

$$\cup \{(x_1, x_2, x_3, x_4, x_5) | x_i = 2, x_j = 2, i \ne j; i, j = 1, 2, 3, 4, 5\}$$

will enable us to estimate main effects. Adding an extra choice set

$$S_5 = \{(1,1,1,1,1)\} \cup \left\{(x_1, x_2, x_3, x_4, x_5) \middle| \begin{matrix} x_i = 2, x_j = 2, x_k = 1; \\ i \ne j \ne k \ne m; i, j, k, m = 1, 2, 3, 4, 5 \end{matrix} \right\}$$

$$\cup \left\{(x_1, x_2, x_3, x_4, x_5) \middle| \begin{matrix} x_i = 2, x_j = 1;, x = 1, x_m = 1; \\ i \ne j \ne k \ne m; i, j, k, m = 1, 2, \ldots 5 \end{matrix} \right\},$$

the choice sets S_3, S_4, S_5 allow estimation of main effects and two-way interactions. Finally, adding

$$S_6 = \{(x_1, x_2, x_3, x_4, x_5) | x_i = 2, x_j = 2, x_k = 2; i \ne j \ne k; i, j, k = 1, 2, 3, 4, 5\}$$

$$\cup \{(x_1, x_2, x_3, x_4, x_5) | x_i = 2, x_j = 2, x_k = 1, x_m = 1; i \ne j \ne k \ne m; i, j, m = 1, 2, \ldots, 5\}$$

$$\cup \left\{(x_i, x_j, x_k, x_m, x_p) \middle| \begin{matrix} x_i = 2, x_j = 1, x_k = 1, x_m = 1, x_p = 1; \\ i \ne j \ne k \ne m \ne p; i, j, k, m, p = 1, 2, 3, 4, 5 \end{matrix} \right\}$$

gives choice sets S_3, S_4, S_5, S_6, which enables us to estimate all main effects, two- and three-attribute interactions. Note that the choice set S_ℓ for $\ell = 3$,

4, 5, 6 has elements $(x_1, x_2, x_3, x_4, x_5)$ such that $\sum_{i=1}^{5} x_i = \ell$. The choice sets S_3, S_4, S_5, S_6 have too many profiles for a respondent to make a meaningful evaluation of the profiles. Hence, we subdivide each of S_3 and S_4 into more choice sets at the first stage, S_5 into subsets at the second stage, and S_6 into subsets at the final stage.

In a sequential experiment, we want to control the probability of selecting at least one main effect or two-factor or three-factor interaction to be significant when all those are zero at the level of significance α, which is usually .05. Let α_1 be the level of significance to test the lack of fit at the first two stages and let α_2 be the level to test each of the main effects and two- or three-factor interactions. The sequential experimentation strategy consists of

1. At the first stage, collect data on choice sets S_ℓ and $S_{\ell+1}$, where $3(s-2) \le l \le (m-3)(s-1)$. Analyze the data and test lack of fit at a significance level α_1. If lack of fit is not significant, test each main effect at the α_2 level of significance. If the lack of fit is significant, proceed to the second stage.
2. Collect data with the choice set $S_{\ell+2}$. Use the choice sets $S_\ell, S_{\ell+1}, S_{\ell+2}$ and fit a model of main effects and two-factor interactions. If there is no lack of fit at the α_1 level, test each main effect and two-factor interaction at the α_2 level. If the lack of fit is significant, proceed to the third stage, which is also the final stage.
3. Collect data with the choice set $S_{\ell+3}$. Using the data from $S_\ell, S_{\ell+1}, S_{\ell+2}$, and $S_{\ell+3}$ test all main effects, two-and three-attribute interactions at the α_2 level.

If **Y** is a response vector with **X** as the design matrix, β as a parameter vector, and **S** as estimated dispersion matrix, the lack of fit sum of squares is

$$Y'S^{-1}Y - \hat{\beta}'(X'S^{-1}X)\hat{\beta},$$

which is asymptotically distributed as a x^2 with $n-r$ degrees of freedom, where n is the dimensionality of **Y**, r is the rank of **X**, and $\hat{\beta} = (\mathbf{X'S^{-1}X})^- (\mathbf{X'S^{-1}Y})$ is the weighted least squares estimator. Recall that A^- is a generalized inverse of A satisfying $AA^-A = A$. The described sequential procedure and the overall significance level α result in the following equation:

$$(1-\alpha_1)\left\{1-(1-\alpha_2)^{m\ main}\right\}+\alpha_1(1-\alpha_1)\left\{1-(1-\alpha_2)^{m\ main+m\ twoway}\right\}$$
$$+\alpha_1^2\left\{1-(1-\alpha_2)^{m\ main+m\ twoway+m\ threeway}\right\}=\alpha.$$

For simplicity, the tests at different stages are assumed independent in this chapter.

Since we may be interested in drawing inferences on main effects, two- and three-factor interactions, we like to choose the level of significance for testing the lack of fit to be high, like $\alpha_1 = 0.2$ or 0.3. We choose $\alpha_1 = 0.2$. The equation then has only one unknown α_2, and we solve for it when $\alpha = .05$. Various α_2 values are given in Table 7.1.

From Table 7.1, in a 4^5 experiment, we test each of the main effects, two- and three-attribute interactions at all stages using the level $\alpha_2 = .001$ corresponding to the entry in line 4 and column headed by 5.

Testing main effects, two- and three-way interaction will involve several tests of hypotheses. Multiplicity of tests is a real problem, and we may use the Boniferroni method and test the contrasts at a level (α_2/number of tests) controlling the family-wise error rates (FWERs), or we may use the Benjamini-Hochberg (1995) method for controlling the false discovery rate (FDR).

Often, in experiments many factors may not have significant main effects, and as the described sequential method is elaborate, we do a screening experiment using only two-level factors to determine the attributes for the larger experiment. Suppose there are t factors where t is a prime or prime power such that $t \equiv 1 \pmod 4$ or $t \equiv 3 \pmod 4$. We initially conduct a screening 2^t experiment using the choice sets as given in Theorem 5.1. The significant m main effects determined at this screening stage will be used to do the earlier sequential experiment using s levels of each factor to determine the significant main effects and interactions. At the screening stage, it is preferable to use a higher level of significance like .2 so borderline significant factors for the elaborate experiment are not missed. This procedure is analogous to using Hadamard matrices as main-effects screening experiments in design of experiments (see the work of Dean and Lewis, 2006; Dean, 2007). Since a Hadamard matrix may give a single-item PO choice set, we suggest the method of Theorem 5.1 for a screening experiment.

TABLE 7.1

α_2 Values with $\alpha_1 = .2$ and $\alpha = .05$

	Number of Attributes					
Level	**2**	**3**	**4**	**5**	**6**	**7**
2	.023	.014	.010	.007	.005	.004
3	.011	.006	.004	.003	.002	.001
4	.007	.003	.002	.001	.001	.001
5	.005	.002	.001	.001	.000	.000

7.3 Sequential Methods to Estimate Main Effects and Interactions, Including a Common Attribute in 2^m Experiments

Consider a 2^m experiment with m attributes $A_1, A_2, \ldots, A_m$. Chen and Chitturi (2008) show that the three choice sets

$$S_2^* = \{100\ldots0,\ 010\ldots000,\ 001\ldots000,\ 000\ldots001\},$$

$$S_2^* = S_2 - \{(000\ldots110)\} = \left\{(x_1, x_2, \ldots, x_n)\middle|\sum_{i=1}^{n} x_i = 2; x_{n-2} \neq 1, x_{n-1} \neq 1\right\}, \text{ and}$$

$$S_3^* = \left\{(x_1, x_2, \ldots, x_n)\middle| x_1 = 1; \sum_{1=2}^{n} x_i = 2 : x_{n-2} \neq 1, x_{n-1} \neq 1\right\}$$

are capable of estimating all main effects, all two-factor interactions, and a three-factor interaction, including a common attribute. These estimates use all three choice sets. Note that S_1^* has one profile less than S_2, and S_1^* has $m - 4$ profiles less than S_1. Furthermore, we cannot estimate all main effects using only S_1^* and S_2^*. In view of these considerations, a sequential plan in this context is

1. At the first stage, collect the data using choice sets S_1 and S_2. Test for the lack of fit of these data in a main-effect plan with level of significance α_1. If there is no lack of fit, test all main effects at level α_2, concluding the study. If lack of fit is significant, proceed to stage 2.
2. Collect additional data with choice set S_3^* and test main effects and all two-factor interactions and a three-factor interaction, including a common attribute, say A_1, at significance level α_2 and terminate the study.

By the end of stage 2, we tested all main effects of the attributes $A_1, A_2, \ldots, A_m$; two-factor interactions of A_1 and A_j for $j = 2, 3, \ldots, m$; and the three-factor interaction A_1, A_2, and A_3.

As in Section 7.2, we have the following equation connecting α_1, α_2 and α

$$(1-\alpha_1)\left\{1-(1-\alpha_2)^m\right\}+\alpha_1\left\{1-(1-\alpha_2)^{2m}\right\}=\alpha.$$

By taking $\alpha_1 = .3$ and $\alpha = .05$, α_2 is $\alpha_2 = 1-(0.91)^{1/m}$. Note that real number α_2 does not exist with $\alpha_1 = .2$.

In a 2^5 experiment, in the first stage data are collected on the two choice sets
$S_1 = \{10000, 01000, 00100, 00010, 00001\}$, and
$S_2 = \{11000, 10100, 10010, 10001, 01100, 01010, 01001, 00110, 00101, 00011\}$.
In the second stage, data are collected on the choice set

$$S_3^* = \{11100, 11010, 11001, 10101, 10011\}.$$

The lack of fit is tested with level $\alpha_1 = .3$, and each contrast is tested with level

$$\alpha_2 = .0187.$$

7.4 CA Testing Main Effects and a Two-Factor Interaction Sequentially

Consider a conjoint study providing each profile sequentially and asking the respondents to rate the profile using (for example) a 1 (least) to 10 (best) rating scale. Each profile is evaluated by N respondents.

Let $A_1, A_2, \ldots, A_m$ be the m, two-level, attributes under investigation. We estimate and test m main effects and a two-factor interaction of A_1 and A_2. Raghavarao and Wiley (2006) prove that the profiles (100 … 00), (000 … 01), and (110 … 00) and the profiles $(x_1, x_2, \ldots, x_m)$ with $x_m = 1$ and $\Sigma_{i=1}^{m-1} x_i = 1$ will enable us to estimate the m main effects and the two-attribute interaction of A_1A_2. We can sequentially present the profiles as follows, and we indicate the profiles that will enable us to estimate and test the contrasts; the tests used here are standard two-sample t tests:

Order of Presenting	Profiles	Contrast Name and Required Profiles
1	000 … 01	
2	100 … 01	2 vs. 1 – A_1 effect
3	010 … 01	3 vs. 1 – A_2 effect
•	•	•
m	000 … 11	m vs. 1 – A_{m-1} effect
$m + 1$	100 … 00	$m + 1$ vs. 2 – A_m effect
$m + 2$	110 … 00	$\{1, m + 2\}$ vs. $\{3, m + 1\}$ –A_1A_2 interaction

7.5 Interim Analysis

Let us consider a main-effects s^m design using two choice sets S_ℓ and $S_{\ell+1}$, where $(s-2) \le \ell \le (m-1)(s-1)$, and each of the two choice sets is evaluated by N respondents. At the end of the investigation, if the analysis indicates at least one main effect is significant, the study is a success, and if no main effect is significant, the study is a failure. Instead of spending money and time on a failed investigation, the investigator may want to abandon the study at the middle of the investigation. Such an analysis at an intermediary stage of the investigation is known as *interim analysis.* One can perform more than one interim analysis, although we restrict ourselves to one interim analysis here. Interim analysis is commonly used in clinical trials.

For convenience, let us assume that the number N of respondents evaluating each of the two choice sets is even, and $N = 2v$. When v respondents evaluated each of the two choice sets, we perform the interim analysis, and we use the following sequential plan:

1. We perform a global test at α_1 level. If the test is not significant, we stop the experiment. If the test is significant, we proceed to stage 2.
2. We collect data using all N respondents and test each main effect at the α_2 level and draw inferences on all main effects.

The relationship between α_1 and α_2 and the overall error rate α is given by

$$\alpha_1\left\{1-\left(1-\alpha_2\right)^m\right\}=\alpha.$$

For example, taking the overall significance level at $\alpha = .05$ and $\alpha_1 = .20$, with $m = 10$, we get $\alpha_2 = .028$.

7.6 Some Sequential Plans for 3^m Experiments

Raghavarao and Wiley (2006) show that, in a 3^m experiment, the following three choice sets

$$S_1 = \left\{x_1x_2...x_m \mid \sum_{i=1}^{m} x_i = 1\right\},$$

$$S_2^* = \{x_1x_2\ldots x_m \mid x_i = 2, i = 1,\ldots,m\}, \text{ and}$$

$$S_2^{**} = \left\{x_1x_2\ldots x_m \mid \sum_{i=1}^{m-1} x_i = 1, x_m = 1\right\}$$

form a main-effects plan. By adding the two choice sets

$$S_3^* = \{110\ldots001, 200\ldots001, 120\ldots000\} \text{ and}$$

$$S_4^* = \{220\ldots000, 210\ldots001\},$$

the choice sets S_1, S_2^*, S_2^{**}, S_3^*, S_4^* are a main-effect plan that is capable of estimating the two-factor interaction A_1A_2. Note that S_2^*, S_2^{**}, $\subset S_2$, S_3^*, $\subset S_3$, and $S_4^* \subset S_4$. The two choice sets S_2^* and S_2^{**} together as one PO choice set and S_1 form a two-choice set main-effects plan, but for convenience we show three choice sets as a main-effects plan.

A sequential method in this case is

1. Using three choice sets S_1, S_2^*, S_2^{**} collect the data and analyze as a main-effects plan. If the lack of fit is not significant at the α_1 level, test all main effects at the α_2 level. If the lack of fit is significant, go to stage 2.
2. Collect additional data using the choice sets S_3^* and S_4^*. Test all main effects and the two-factor interaction at the α_2 level.

Following Section 7.3, for $\alpha = .05$, we can take $\alpha_1 = .3$, $\alpha_2 = 1-(0.91)^{1/m}$.

A sequential plan for a 3^m experiment can be formed using Theorem 5.6 of Chapter 5.

8

Mixture Designs

8.1 Introduction

In Chapter 6, we discussed the brands of a product with attribute cross-effects and brand-effects models. The objective in that chapter was to determine the effect of sales of a brand when another brand in the market is replaced by a new brand and to determine the choice of brands to maximize the revenue. Our objective in this chapter is determining the optimal allocation of a fixed resource (such as money or research effort) to categories (such as product attributes).

A key aspect of economic environments is the fact that producers and consumers operate under budget constraints. Consumers' demand will be influenced by their preferences for attribute compositions but constrained by the funds they have available to purchase offerings. Conversely, at a given price, the requirement that producers make a profit constrains the costs at which they can produce a product or service. This serves to restrict the feasible attribute compositions they may supply at that price. The profit-maximizing producer faces two separate issues. The first is to determine the most preferred attribute composition for a product at a given cost (price). The second is to determine the profit-maximizing price (cost). Current approaches to conjoint analysis (CA) and discrete choice experimentation (DCE) design (such as fractional factorial designs) confound these two issues. The mixture/amount approach to experimental design discussed in this chapter provides a means for untangling these issues. We introduced mixture designs in Chapter 2. These designs are widely used in industrial experiments, and their use in CA and DCE is demonstrated in the work of Raghavarao and Wiley (2009). Piepel and Cornell (1985) discuss mixture amount designs in industrial settings.

8.2 Mixture Designs: CA Example

To motivate the difference between the designs discussed in previous chapters and those discussed in this chapter, we begin with mixture design applied to a CA study. Let us consider a mixture design with m attributes and profiles $(x_1, x_2, \ldots, x_m)$ satisfying $0 \leq x_i \leq 1$, $\sum x_i = 1$, where in contrast to previous chapters, the attributes are on a continuous scale. We noted in Chapter 2 that the response model for a mixture design consists of linear and cross-product terms of x_i's with no intercept and no square terms. The model consists of $m(m + 1)/2$ parameters, and to estimate all the parameters, the minimal design consists of $m(m + 1)/2$ profiles. We require designs with "design points" sufficient to estimate all linear and cross-effects parameters.

Simplex centroid and simplex lattice are two designs used to provide design points in mixture experiments. With a simplex-centroid design, the design points correspond to m permutations of (1, 0, 0, … , 0), $\binom{m}{2}$ permutations of (.5, .5, 0, … , 0), $\binom{m}{3}$ permutations of (1/3, 1/3, 1/3, 0, … , 0), … , and finally the single $(1/m, 1/m, \ldots, 1/m)$ mixture. The number of distinct points is $2^m - 1$, so a {4, 4} simplex centroid has 15 design points, which consist 100% of discretionary resources allocated to each of m attributes, 50% allocated to each pair of attributes, and so forth. Note that all design points need not be used. With $m = 4$, there are 10 parameters for the quadratic model with cross products, so a minimum of 10 of the 15 available design points may be selected.

With an $\{m, n\}$ simplex-lattice design, the proportions assigned to m attributes take $n + 1$ equally spaced values from 0 to 1, $x_i = 0, 1/n, 2/n, \ldots, 1$ for $i = 1, 2, \ldots, m$, with $\sum x_i = 1$. There are $(m + n - 1)!/(n!(m - 1)!)$ design points in the $\{m, n\}$ simplex lattice. With a {4, 4} simplex-lattice design, there are 35 available design points, and a minimum of 10 are required to estimate the parameters of a linear component model with cross products. The examples in this chapter are simplex-lattice designs. Discussions of lattice designs in mixture experiments may be found in the work of Scheffé (1958, 1963) and Cornell (1990).

A simplex-lattice design $\{m, 2\}$ consists of $m(m + 1)/2$ profiles, and we can estimate all the parameters of the quadratic model with cross products. However, these profiles have the drawback that each profile has only one or two nonzero levels. Although the design consisting of at least one profile with equal nonzero levels for each of the m attributes does not have statistical optimality properties, it is intuitively appealing, and respondents have the opportunity to evaluate such a profile $(1/m, 1/m, \ldots, 1/m)$. To include the profile $(1/m, 1/m, \ldots, 1/m)$, we can consider a simplex-lattice design $\{m, m\}$ for experimentation. A simplex-lattice $\{m, m\}$ design will have many more profiles than needed, and the experimenter can choose suitable profiles from simplex-lattice $\{m, m\}$ design to estimate parameters and provide opportunities to the respondents to make a choice by comparing the profiles.

For example, suppose a restaurant owner wishes to arrange a lunch plate with a fixed cost. A decision has to be made for how much of that cost goes to each of the menu items: appetizers, main course, dessert, and drinks. The manager may also be interested to find the total cost and splitting the cost on each of the four food items the patrons of that restaurant like. This second problem is called a mixture/amount design. In the following sections, we discuss these problems in some detail.

Let the "price point" of the manager be \$15.00. After overhead, \$9.00 is available to spend on a meal consisting of a drink, appetizer, main course, and dessert. Since all the four items must be included in the plate, the manager wants a minimum of \$0.75 for drink, \$0.50 for appetizer, \$1.50 for main course, and \$0.65 for dessert. This leaves a discretionary amount of \$5.60 to be used for the four food items, and the objective is to determine the optimum allocation of the \$5.60 to the four food items.

Let a CA be performed using 13 profiles given in Table 8.1. Table 8.1 gives the mixture profile, the actual amount for each item, and the average response. For example, with the first profile, 100% of the discretionary budget is allocated to drinks and the minimum permissible allocation to the remaining categories. For the fifth profile, one-half the discretionary budget is allocated to drinks, one-quarter to appetizers, and one-quarter to the main course. The profiles are given to 100 respondents who rate the profiles using a 1–10 scale.

The fitted response equation for the response Y on the mixture profile, (x_1, x_2, x_3, x_4) is

$$\hat{Y} = .61x_1 - .59x_2 - 1.34x_3 + .48x_4 - .61x_1x_2 + .61x_1x_3 + .19x_1x_4 + 1.54x_2x_3 - .51x_2x_4 + .36x_3x_4. \tag{8.1}$$

Differentiating $\hat{Y}$ with respect to x_1, x_2, x_3, x_4 such that $\sum x_i = 1$ and equating to zero, we get the critical values x_1^*, x_2^*, x_3^* and x_4^* of 0%, 55%, 45%, and 0%, respectively. The optimum allocations preferred by the respondents are drink (\$0.75 + \$5.60x_1^*), appetizer (\$0.50 + \$5.60x_2^*), main course (\$1.50 + \$5.60x_3^*), dessert (\$0.65 + \$5.60x_4^*), for a total of \$9.00.

TABLE 8.1

Profiles, Experimental Design, and Average Response

	Mixture				Actual Amount				
Profile	Drink	App.	Main	Dessert	Drink	App.	Main	Dessert	Average Response
1	100%	0%	0%	0%	\$6.35	\$0.50	\$1.50	\$0.65	8.00
2	0%	100%	0%	0%	\$0.75	\$6.10	\$1.50	\$0.65	6.00
3	0%	0%	100%	0%	\$0.75	\$0.50	\$7.10	\$0.65	1.00
4	0%	0%	0%	100%	\$0.75	\$0.50	\$1.50	\$6.25	5.00
5	50%	25%	25%	0%	\$3.55	\$1.90	\$2.90	\$0.65	9.00
6	0%	50%	25%	25%	\$0.75	\$3.30	\$2.90	\$2.05	9.00
7	25%	0%	50%	25%	\$2.15	\$0.50	\$4.30	\$2.05	8.00
8	25%	25%	0%	50%	\$2.15	\$1.90	\$1.50	\$3.45	4.00
9	25%	50%	25%	0%	\$2.15	\$3.30	\$2.90	\$0.65	10.00
10	0%	25%	50%	25%	\$0.75	\$1.90	\$4.30	\$2.05	10.00
11	25%	0%	25%	50%	\$2.15	\$0.50	\$2.90	\$3.45	8.00
12	50%	25%	0%	25%	\$3.55	\$1.90	\$1.50	\$2.05	4.00
13	25%	25%	25%	25%	\$2.15	\$1.90	\$2.90	\$2.05	8.00

Instead of using the optimal allocations, suppose the manager decides to divide the discretionary budget percentage-wise equally among the meal components, resulting in drink (\$2.15), appetizer (\$1.90), main course (\$2.90), and dessert (\$2.05). The expected score then is

$$\begin{aligned}\hat{Y} = {} & .61(2.15) - .59(2.90) - 1.34(2.90) + .48(2.05) \\ & - .61\,(2.15)(1.90) + .61(2.15)(2.90) + .19(2.15)(2.05) \\ & + 1.54(1.90)(2.90) - .51(1.90)(2.05) \\ & + .36(2.90)(2.05) = 8.07.\end{aligned}$$

8.3 Mixture Designs: DCE Example

The set of profiles of a simplex-lattice $\{m, n\}$ design are Pareto optimal; hence, any subset of profiles is also Pareto optimal. To have a DCE study, we need to construct choice sets of selected q profiles, where $q > m(m + 1)/2$. These can be constructed by partitioning the q profiles in suitable choice sets with one replication of each profile; this will not allow the respondents to make pairwise comparison of all profiles. It is possible to prepare a minimal number of choice sets in which each set is of same size, every profile occurs equally often, and every pair of profiles occurs equally often. These requirements lead us to choose the sets of a symmetrical balanced incomplete block (BIB) design with $v = q = b$, $r = k$, λ as choice sets. With $m = 4$, we use $q = 13$ profiles given in Table 8.1 to construct 13 choice sets using the BIB design with parameters $v = 13 = b$, $r = k = 4$, $\lambda = 1$. The 13 choice sets with these 13 profiles are given in Table 8.2.

For example, choice set 1 has meal profiles 1, 2, 4, 10, and no choice. Respondents choose one of these five options, and analysis will be based on the logit of the proportion picking each option in the choice set to the proportion choosing the no-choice option as described throughout the monograph. Note that due to the limited availability of symmetrical BIB designs, in some cases we may have to prepare the choice sets for which not all pairs of profiles occur equally often in choice sets. These designs may be PBIB designs (see the work of Raghavarao and Padgett, 2005), discussed in Section 8.5.

Turning to the restaurant example, we consider the 13 choice sets of the previous section and have (say) 200 respondents choose the most preferred profile from each set. If no profile is of interest, they may choose the no-choice option. In Table 8.3, we indicate the mixtures in the sets, proportion choosing the profile.

We need to perform weighted least squares to estimate the parameters. The 52-dimensional response vector $\mathbf{Y}$ is

TABLE 8.2

Balanced Incomplete Block Design

Choice Set	1	2	3	4	5	6	7	8	9	10	11	12	13
	1	2	3	4	5	6	7	8	9	10	11	12	13
	2	3	4	5	6	7	8	9	10	11	12	13	1
Contents	4	5	6	7	8	9	10	11	12	13	1	2	3
	10	11	12	13	1	2	3	4	5	6	7	8	9
	No choice	No choice	No choice	No choice	No choice	No choice	No choice	No choice	No choice	No choice	No choice	No choice	No choice

TABLE 8.3
Discrete Choice Experiment for $m = 4$ in 13 Choice Sets

Choice Set	Mixture Proportions				
1	(1, 0, 0, 0)	(0, 1, 0, 0)	(0, 0, 0, 1)	(0, 1/4, 1/2, 1/4)	NC
	.24	.19	.18	.28	.11
2	(0, 1, 0, 0)	(0, 0, 1, 0)	(1/2, 1/4, 1/4,0)	(1/4, 0, 1/4, 1/2)	NC
	.20	.13	.28	.27	.12
3	(0, 0, 1, 0)	(0, 0, 0, 1)	(0, 1/2, 1/4, 1/4)	(1/2, 1/4, 0, 1/4)	NC
	.14	.22	.32	.19	.13
4	(0, 0, 0, 1)	(1/2, 1/4, 1/4, 0)	(1/4, 0, 1/2, 1/4)	(1/4, 1/4, 1/4, 1/4)	NC
	.18	.25	.23	.23	.10
5	(1/2, 1/4, 1/4, 0)	(0, 1/2, 1/4, 1/4)	(1/4, 1/4, 0, 1/2)	(1, 0, 0, 0)	NC
	.25	.26	.15	.23	.11
6	(0, 1/2, 1/4, 1/4)	(1/4, 0, 1/2, 1/4)	(1/4, 1/2, 1/4, 0)	(0, 1, 0, 0)	NC
	.25	.22	.26	.17	.10
7	(1/4, 0, 1/2, 1/4)	(1/4, 1/4, 0, 1/2)	(0, 1/4, 1/2, 1/4)	(0, 0, 1, 0)	NC
	.27	.17	.31	.13	.12
8	(1/4, 1/4, 0, 1/2)	(1/4, 1/2, 1/4, 0)	(1/4, 0, 1/4, 1/2)	(0, 0, 0, 1)	NC
	.16	.29	.25	.19	.11
9	(1/4, 1/2, 1/4, 0)	(0, 1/4, 1/2, 1/4)	(1/2, 1/4, 0, 1/4)	(1/2, 1/4, 1/4, 0)	NC
	.26	.26	.15	.23	.10
10	(0, 1/4, 1/2, 1/4)	(1/4, 0, 1/4, 1/2)	(1/4, 1/4, 1/4, 01/4)	(0, 1/2, 1/4, 1/4)	NC
	.25	.22	.21	.23	.09
11	(1/4, 0, 1/4, 1/2)	(1/2, 1/4, 0, 1/4)	(1, 0, 0, 0)	(1/4, 0, 1/2, 1/4)	NC
	.25	.16	.24	.24	.11
12	(1/2, 1/4, 0, 1/4)	(1/4, 1/4, 1/4, 1/4)	(0, 1, 0, 0)	(1/4, 1/4, 0, 1/2)	NC
	.19	.28	.22	.18	.13
13	(1/4, 1/4, 1/4, 1/4)	(1, 0, 0, 0)	(0, 0, 1, 0)	(1/4, 1/2, 1/4, 0)	NC
	.24	.24	.12	.29	.11

$$Y' = \left(\ln\left(\frac{.24}{.11}\right), \ln\left(\frac{.19}{.11}\right), \ln\left(\frac{.18}{.11}\right), \ln\left(\frac{.28}{.11}\right), \ldots, \ln\left(\frac{.24}{.11}\right), \ln\left(\frac{.24}{.11}\right), \ln\left(\frac{.12}{.11}\right), \ln\left(\frac{.29}{.11}\right)\right).$$

Using the results of Chapter 2, the 4 × 4 estimated dispersion matrix for the first choice set is

$$\sum_1 = \begin{bmatrix} .073 & .050 & .050 & .050 \\ .050 & .077 & .050 & .050 \\ .050 & .050 & .076 & .050 \\ .050 & .050 & .050 & .067 \end{bmatrix}.$$

The 52 × 10 design matrix **X** is given in Table 8.4.

Let β be a 10 × 1 column vector of parameters given by

$$\beta' = (a_1, a_2, a_3, a_4, a_{12}, a_{13}, a_{14}, a_{23}, a_{24}, a_{34}).$$

Let Σ be a 52 × 52 dispersion matrix given by the following diagonal blocks of matrices:

$$\Sigma = D(\Sigma_1, \Sigma_2, \ldots, \Sigma_{13}).$$

The weighted normal equations provide the estimated parameters:

$$\hat{\beta} = (X\Sigma^{-1}X)^{-1} X'\Sigma^{-1}Y.$$

Using these estimated parameters, $\hat{\mathbf{Y}}$ is obtained, and $\hat{\mathbf{Y}}$ is maximized with respect to the mixture component (x_1, x_2, x_3, x_4) with $\Sigma x_i = 1$.

TABLE 8.4

Design Matrix for Mixture Experiment

X =	1	0	0	0	0	0	0	0	0	0
	0	1	0	0	0	0	0	0	0	0
	0	0	0	1	0	0	0	0	0	0
	0	1/4	1/2	1/4	0	0	0	1/8	1/16	1/8
					…					
	1/4	1/4	1/4	1/4	1/16	1/16	1/16	1/16	1/16	1/16
	1	0	0	0	0	0	0	0	0	0
	0	0	1	0	0	0	0	0	0	0
	1/4	1/2	1/4	0	1/8	1/16	0	1/8	0	0

8.4 Mixture–Amount Designs

Price plays an important role in a purchase decision. In Section 8.3, we discussed the optimal mix of the attributes in a profile for a given price. It is very likely that the purchase decision may depend on the price, and for different prices, the mix of attributes may change.

We consider the same problem as in Section 8.3. A restaurant manager wishes to provide a lunch plate for \$15 with \$9 discretionary amount to be assigned to the four food items. The same minimum expenditures to categories apply as in Section 8.3. Suppose the manager wants to determine the best price acceptable to the patrons and the mix of the attributes at that price. Let there exist a symmetrical BIB design with $v = b$, $r = k$, $\lambda = 1$, where v is at least $m(m + 1)/2$ the number of parameters in the mixture design and less than or equal to the maximum number of points in a simplex lattice $\{m, m\}$. In this case, we consider k price levels with the intention of having each price level appear in each choice set. If profiles having 100% allocation to single attributes are used, and if x_{max} (<1) is the maximum level of attributes in the profiles other than the 100% allocation, we need to have $u_{min} > (x_{max})(u_{max})$, where u_{min} and u_{max} are the minimum and maximum prices in the study, respectively. This inequality is needed to make the choice sets Pareto optimal; otherwise, the profile with a single attribute with minimum cost will dominate other profiles and is selected more often.

In the restaurant example, let $u_1 = u_{min} = \$5.60$ amount available after minimum allocations to food categories are satisfied, and we have an existing symmetrical BIB design with parameters $v = 13 = b$, $r = 4 = k$, $\lambda = 1$. We use the 13 profiles 1–13 given in Section 8.3. Among these 13 profiles, we have the single-attribute profiles that have 100% allocation to a single attribute. Hence, the discretionary price u_{max} must satisfy $u_{max} < u_{min}/(1/2)$ as $x_{max} = 1/2$. We choose $u_{max} = \$10.60$. Let us choose two more discretionary amount levels and let $u_{min} = u_1 = \$5.60$, $u_2 = \$7.60$, $u_3 = \$9.60$, $u_4 = u_{max} = \$10.60$. These discretionary amounts lead to prices of \$15, \$17, \$19, and \$20, respectively. We denote by $u_i(\ell)$ the ℓth profile with discretionary amount u_i. For example, u_2 (7) = u_2 (1/4, 0, 1/2, 1/4) denotes the profile with \$(0.75 + 1/4 (7.60)) = \$2.65 on drinks, \$0.50 on appetizers, \$(1.50 + 1/2 (7.60)) = \$5.30 on main course, and \$(0.65 + 1/4 (7.60)) = \$2.55 on desserts.

The response Y on a point scale in CA or natural logarithm of the proportion choosing the profile to the proportion choosing no choice in choice-based conjoint analysis (CBC) will be modeled as

$$E(y) = \alpha u + \beta u^2 + \sum_{i=1}^{m-1} \gamma_i u x_i + \sum_{i=1}^{m} \delta_i x_i + \sum_{\substack{i=1 \\ i<j}}^{m} \delta_{ij} x_i x_j \tag{8.2}$$

for the profile $u(x_1, x_{2, \ldots,} x_m)$. This model is motivated to find the amount u that maximizes (or minimizes) preference or choice in a way that captures the interaction between price and the attributes. In Equation 8.2, we have not used the term $\gamma_m x_m$ because $\sum_{i=1}^{m} x_i = 1$. We emphasize that the model is descriptive rather than explanatory. The objective is to describe changes in the dependent variable as levels of the attributes change with the aim of identifying optimum profiles subject to a budget constraint. The 13 choice sets are given in Table 8.5. In that table, we give artificial data on the proportion choosing the profile and the natural logarithm of this proportion to the proportion choosing the no-choice option.

In the amount–mixture choice sets given in Table 8.6, every two profiles with two different discretionary amount levels occur together. To capture

TABLE 8.5

DCE of Amount–Mixture Design with $n = 4$ and Artificial Data

	Proportions/Logits				
Choice Set	u_1	u_2	u_3	u_4	nc
1	0.17	0.18	0.24	0.29	0.12
	0.35	0.44	0.71	0.91	
2	0.20	0.17	0.21	0.28	0.13
	0.41	0.25	0.46	0.72	
3	0.18	0.25	0.29	0.15	0.13
	0.27	0.63	0.78	0.10	
4	0.21	0.20	0.23	0.23	0.13
	0.51	0.47	0.62	0.61	
5	0.21	0.26	0.19	0.21	0.14
	0.46	0.39	0.84	0.64	
6	0.19	0.20	0.22	0.27	0.12
	0.38	0.53	0.69	0.28	
7	0.19	0.13	0.28	0.23	0.12
	0.48	0.69	0.19	0.68	
8	0.19	0.24	0.26	0.17	0.13
	0.52	0.61	0.60	0.91	
9	0.21	0.26	0.16	0.25	0.13
	0.49	0.30	0.69	0.72	
10	0.19	0.21	0.21	0.28	0.11
	0.52	0.61	0.60	0.91	
11	0.20	0.17	0.25	0.26	0.12
	0.49	0.30	0.69	0.72	
12	0.22	0.27	0.19	0.17	0.16
	0.34	0.55	0.19	0.10	
13	0.21	0.18	0.24	0.24	0.13
	0.462	0.316	0.601	0.611	

the decisions of respondents in a general way, it is preferable that every pair of profiles with the same price occur in a choice set; further, every pair of prices associated with a profile should occur in a choice set (Table 8.6). Such a design is called a rectangular partially BIB design. We write our 52 amount–mixture profiles in a 13 × 4 array

$u_1(1)$	$u_2(1)$	$u_3(1)$	$u_4(1)$
$u_1(2)$	$u_2(2)$	$u_3(2)$	$u_4(2)$
	…		
$u_1(13)$	$u_2(13)$	$u_3(13)$	$u_4(13)$

and we construct a block design in which every pair of amount–mixtures occurring in the same row (column) is given in λ_1 (λ_2) sets, and every other mixture–amount profiles occurs in λ_3 sets. This design can be considered as a Kronecker product of two BIB designs. If the researcher is unwilling to include different amount levels for the same profile in a choice set, then we set $\lambda_1 = 0$, and the design is a group-divisible design.

The sets (u_1, u_2); (u_1, u_3); (u_1, u_4); (u_2, u_3); (u_2, u_4); (u_3, u_4) form a BIB design with the price levels. The sets (1, 2, 4, 10); (2, 3, 5, 11); (3, 4, 6, 12), (4, 5, 7, 13); (5, 6, 8, 1); (6, 7, 9, 2); (7, 8, 10, 3); (8, 9, 11, 4); (9, 10, 12, 5); (10, 11, 13, 6); (11, 12, 1, 7); (12, 13, 2, 8); (13, 1, 3, 9) form a BIB design with the profiles.

The Kronecker product of the set (u_1, u_2) of the first design and the set (1, 2, 4, 10) of the second design is a set of size 8 given by

$$\{u_1(1), u_1(2), u_1(4), u_1(10), u_2(1)\ u_2(2), u_2(4), u_2(10)\}.$$

The (6)(13) = 78 sets of size 8 by taking the Kronecker product of the sets of the first BIB design and the sets of the second BIB design are a rectangular partially BIB design, and they are more ideal mixture–amount choice sets than the design given in Table 8.5.

Returning to the responses given in Table 8.5, let

$$Y' = (0.35, 0.44, 0.71, 0.91, \ldots, 0.462, 0.316, 0.601, 0.611).$$

Let $\Sigma_1, \Sigma_2, \ldots, \Sigma_{13}$ be the dispersion matrices of the responses given in the 13 choice sets, where

$$\Sigma_1 = D(1/.3, 1/.4, 1/.1, 1/.1) + (1/.1)J_4, \ldots, \Sigma_{13} = D(1/.2, 1/.3, 1/.2, 1/.2) + (1/.2)J_4.$$

Let **X** be the 52 × 13 design matrix

TABLE 8.6

Price Levels ($) in Design

Choice	$15 Meal				$17 Meal				$19 Meal				$20 Meal			
Set	Drink	App.	Main	Dessert	Drink	App.	Main	Dessert	Drink	App.	Main	Dessert	Drink	App.	Main	Dessert
1	6.35	0.50	1.50	0.65	0.75	8.10	1.50	0.65	0.75	0.50	1.50	10.25	0.75	3.15	6.80	3.30
2	0.75	6.10	1.50	0.65	0.75	0.50	9.10	0.65	5.55	2.90	3.90	0.65	3.40	0.50	4.15	5.95
3	0.75	0.50	7.10	0.65	0.75	0.50	1.50	8.25	0.75	5.30	3.90	3.05	6.05	3.15	1.50	3.30
4	0.75	0.50	1.50	6.25	4.55	2.40	3.40	0.65	3.15	0.50	6.30	3.05	3.40	3.15	4.15	3.30
5	3.55	1.90	2.90	0.65	0.75	4.30	3.40	2.55	3.15	2.90	1.50	5.45	6.05	3.15	4.15	0.65
6	0.75	3.30	2.90	2.05	2.65	0.50	5.30	2.55	3.15	5.30	3.90	0.65	0.75	5.80	4.15	3.30
7	2.15	0.50	4.30	2.05	2.65	2.40	1.50	4.45	0.75	2.90	6.30	3.05	3.40	0.50	6.80	3.30
8	2.15	1.90	1.50	3.45	2.65	4.30	3.40	0.65	3.15	0.50	3.90	5.45	3.40	3.15	1.50	5.95
9	2.15	3.30	2.90	0.65	0.75	2.40	5.30	2.55	5.55	2.90	1.50	3.05	3.40	5.80	4.15	0.65
10	0.75	1.90	4.30	2.05	2.65	0.50	3.40	4.45	3.15	2.90	3.90	3.05	0.75	3.15	6.80	3.30
11	2.15	0.50	2.90	3.45	4.55	2.40	1.50	2.55	10.35	0.50	1.50	0.65	3.40	0.50	4.15	5.95
12	3.55	1.90	1.50	2.05	2.65	2.40	3.40	2.55	0.75	10.10	1.50	0.65	6.05	3.15	1.50	3.30
13	2.15	1.90	2.90	2.05	8.35	0.50	1.50	0.65	0.75	0.50	11.10	0.65	3.40	3.15	4.15	3.30

$$X = \begin{pmatrix} u_i & u_1^2 & (1)u_1 & (0\ u_1) & (0\ u_1) & 1 & 0 & 0 & 0 & 0 & 0 & 0 & 0 & 0 & 0 \\ & & & & \cdots & & & & & & & & & & \\ u_4 & u_{42} & (1/4)\ u_4 & (1/2)u_4 & (1/4)u_4. & 1/4 & 1/2 & 1/4 & 0 & 1/8 & 1/16 & 0 & 1/8 & 0 & 0 \end{pmatrix}.$$

Let θ be the 13-component vector of parameters given by

$$\theta' = (\alpha, \beta, \dots, \delta_{34}).$$

Letting $\Sigma = D(\Sigma_1, \Sigma_2, \dots, \Sigma_3)$, the weighted least squares estimator $\hat{\theta}$ of θ is then $\hat{\beta} = (X'\Sigma^{-1}X)^{-1}X'\Sigma^{-1}Y)$. Using $\hat{\theta}$, we get the estimated Y, and maximizing it, we determine optimal u and x_1, x_2, x_3, x_4.

8.5 Other Mixture Designs

With $m = 3$, in a simplex-lattice (3, 3) design, the parameters are all estimable using the following $m = 6$ profiles:

a. (1, 0, 0); b. (0, 1, 0); c. (0, 0, 1); d. (1/3, 0, 2/3); e. (1/3, 2/3, 0); f. (1/3, 1/3, 1/3).

With these profiles, we can form three choice sets:

Choice Sets	Contents
1	{a, d, b, e, no choice}
2	{a, d, c, f, no choice}
3	{b, e, c, f, no choice}

in which the three pairs of profiles (a, d), (b, e), (c, f) occur in two choice sets, while every other pair of profiles occurs once in the choice sets. The design used to form such choice sets is known as group-divisible design.

Alternatively, one can use the six profiles in four choice sets forming a group-divisible design of three alternatives as follows:

Choice Sets	Contents
1	{a, b, c, no choice}
2	{a, e, f, no choice}
3	{d, b, f, no choice}
4	{d, e, c, no choice}

In these four choice sets, the profile pairs (a, d); (b, e); (c, f) do not occur together in any choice set, whereas every other pair of profiles occurs together in exactly one choice set.

With $m = 3$, if we use seven profiles instead of the six profiles discussed as given by I. (1, 0, 0); II.(0, 1, 0); III. (0, 0, 1); IV. (2/3, 1/3, 0); V. (0, 2/3, 1/3); VI. (1/3, 0, 2/3); VII. (1/3, 1/3, 1/3), we can use a BIB design with parameters $v = 7 = b$, $r = 3 = k$, $\lambda = 1$ to form the following seven choice sets:

Choice Set	Contents
1	{I, II,IV, no choice}
2	{II, III, V, no choice}
3	{III, IV, VI, no choice}
4	{IV, V, VII, no choice}
5	{V, VI, I, no choice}
6	{VI, VII, II, choice}
7	{VII, I, III, no choice}

8.6 Mixture Designs: Field Study Illustration

We now illustrate mixture designs in the context of a field experiment. Suppose the manager of a supermarket chain has 24 shelf spaces to allocate to three suppliers A, B, and C of a product and wants to find the proportion of the shelf space to be allocated to the three suppliers that would maximize the revenue of that space. The assumption is that, given an allocation of shelf space, each supplier allocates the proportion of its own brands to the available number of facings that maximizes its own profit.

To solve the allocation problem at the supplier level, the manager uses a mixture design to determine the optimal proportions of allocated shelf space to maximize the revenue of the total space (24 facings). The manager takes 10 stores and uses the design points in Table 8.7. Facings are assigned to suppliers in each of the 10 stores to match the design. Table 8.7 shows the fractional and actual number of brands in each of the 10 stores and artificial revenue generated on that product in 12 weeks.

We fit the linear model

$$E(Y) = \sum_{i=1}^{3} \alpha_i x_i + \sum_{\substack{i>1 \\ i,j=1}}^{3} \alpha_{ij} x_i x_j$$

where Y is the response from the mixture (x_1, x_2, x_3) with $x_1 + x_2 + x_3 = 1$. As indicated in Chapter 2, the model is free of a constant term and square terms.

The fitted model is

$$\hat{Y} = 262.4x_1 + 342x_2 + 340.3x_3 + 104.5x_1x_2 + 277.9x_1x_3 - 49x_2x_3. \quad (8.3)$$

Using Equation 8.3, the expected revenue in a week for selling that product storing 12 A's, 8 B's, and 4 C's is

$$\hat{Y} = (262.4)(1/2) + (342^2)(1/3) + (340.3)(1/6) + (104.5)(1/2)(1/3) + 277.9(1/2)(1/6) - (49)(1/3)(1/6) = \$329.76$$

Maximizing $\hat{Y}$ given in Equation 8.3 subject to the condition $x_1 + x_2 + x_3 = 1$, we get

$$x_1 = 0.10, x_2 = 0.85, x_3 = 0.05.$$

TABLE 8.7

Number of Items of the Brands and Artificial Revenue

	No. of Items			Fraction			
Store	**A**	**B**	**C**	**A**	**B**	**C**	**Revenue**
1	24	0	0	1	0	0	$300
2	0	24	0	0	1	0	$350
3	0	0	24	0	0	1	$320
4	16	8	0	2/3	1/3	0	$270
5	8	16	0	1/3	2/3	0	$355
6	16	0	8	2/3	0	1/3	$280
7	8	0	16	1/3	0	2/3	$420
8	0	16	8	0	2/3	1/3	$290
9	0	8	16	0	1/3	2/3	$345
10	8	8	8	1/3	1/3	1/3	$400

TABLE 8.8

Expected Weekly Revenues at Optimal or Near-Optimal Mix of the Brands

No. of Mixture Amounts			Expected Weekly Revenue
A	**B**	**C**	
0.10	0.86	0.04	$342.28
0.12	0.84	0.03	$342.75
0.08	0.87	0.06	$341.60
0.09	0.88	0.03	$342.41
0.11	0.83	0.06	$342.18
0.09	0.84	0.07	$341.54
0.11	0.87	0.02	$342.92

This implies that we maximize the revenues selling that product by allocating facings to suppliers in .10, .85, and .05 proportions. With 24 items, we put 2, 20, and 1 item (round-off adjustment) of the three suppliers A, B, and C to maximize the revenue. We like to know the expected revenues near the optimal mix, and these revenues are given in Table 8.8.

The mixture design approach described presumes that the manager has settled on 24 as the number of shelf facings to give to the three suppliers of that brand. Perhaps this number is given structurally, as would be the case, for example, if 24 facings were the number of facings on one aisle of a store. However, it may be the case that the number of facings to allocate to the brand is itself a variable, and the manager may also want to determine the optimal number of facings to allocate. This is a mixture–amount problem. In mixture–amount designs, both the amount of a resource to allocate to categories and the optimal allocation of the resource among categories may be experimentally determined. Mixture–amount designs are discussed in Section 8.4.

References

Adamowicz, W. P., M. Boxall, and M. Williams. 1998. Stated preference approaches for measuring passive use values: Choice experiments versus contingent valuation. *American Journal of Agricultural Economics* 80:64–75.

Anderson, D. A., W. J. Borgers, D. Ettema, and H. J. P. Timmermans. 1992. Estimating availability effects in travel choice modeling: A stated choice approach. *Transportation Research Record* 1357:51–65.

Anderson, D. A., and J. B. Wiley. 1992. Efficient choice set designs for estimating availability cross-effects models. *Marketing Letters* 3(4):357–70.

Anderson, N. H. 1981. *Foundations of information integration theory*. New York: Academic Press.

Andrews, R. L., A. Ansari, and I. S. Currim. 2002. Hierarchical Bayes versus finite mixture conjoint analysis models: A comparison of fit, prediction, and part-worth recovery. *Journal of Marketing Research* 39:187–98.

Andrich, D. 1978. A rating formulation for ordered response categories. *Psychometrika* 43(4):561–73.

Bass, F. M., E. A. Pessimier, and D. R. Lehmann. 1972. Experimental study of relationships between attitudes, brand preferences and choice. *Behavioral Science* 17:532–41.

Batsell, R. R., and J. C. Polking. 1985. A generalized model of market share. *Marketing Science* 4:177–98.

Bechtel, G., and J. B. Wiley. 1983. Probabilistic measurement of attributes: A logit analysis by least squares. *Marketing Science* 2(4):389–405.

Becker, G. M., M. H. DeGroot, and J. Marschak. 1963. Probabilities of choice among very similar objects. *Behavioral Science* 8:306–311.

Benjamini, Y., and Y. Hochberg. 1995. Controlling the false discovery rate: A practical and powerful approach to multiple testing. *Journal of the Royal Statistical Society* B(57):289–300.

Bhaumik, D. K. 1995. Optimality in the competing effects model. *Sankhya* B(57):48–56.

Box, G. E. P., and K. B. Wilson. 1951. On the experimental attainment of optimal conditions. *Journal of the Royal Statistical Society* 13B: 1–45.

Box, G. E. P., and K. B. Wilson. 1951. On the experimental attainment of optimum conditions. *Journal of the Royal Statistical Society Series B* 13:1–45.

Brandley, M. A., and H. Gunn. 1990. Stated preference analysis of values of travel time in the Netherlands. *Transportation Research Record* 1285:78–89.

Bullock, C. H., D. A. Elston, and N. A. Chalmers. 1998. An application of economic experiments to a land-use, deer hunting, and landscape change in the Scottish highlands. *Journal of Environmental Management* 52:335–51.

Calvin, L. D. 1954. Doubly balanced incomplete block designs for experiments in which the treatment effects are correlated. *Biometrics* 10(1):61–88.

Cattin, P., and D. R. Wittink. 1982. Commercial use of conjoint analysis: A survey. *Journal of Marketing* 46(Summer):44–53.

Chen, J., and P. Chitturi. 2008. Choice experiments for estimating main effects and two factor interactions. *American Statistical Association Conference Proceedings* Denver.

Chitturi, R., P. Chitturi, and D. Raghavarao. 2010. Forthcoming. Design for synergy with brand or price information. *Psychology and Marketing*.

Churchman, C. W. 1961. *Prediction and optimal decisions: Philosophical issues in a science of values*. Englewood Cliffs, NJ: Prentice-Hall.

Cornell, J. A. 1990. *Experiments with mixtures: Designs, models, and the analysis of mixture data*. New York: Wiley.

Darmon, R. Y., and D. Rouziès. 1999. Internal validity of conjoint analysis under alternative measurement procedures. *Journal of Business Research* 46:67–81.

Dawes, R. M., and B. Corrigan. 1974. Linear models in decision making. *Psychological Review* 81:95–106.

Dean, A. 2007. Screening experiments with large numbers of factors. Paper presented at the International Conference on Advances in Interdisciplinary Statistics and Combinatories, October 12–14, Greensboro, NC.

Dean, A., and S. Lewis. 2006. *Screening: Methods, for experimentation in industry, drug discovery, and genetics*. New York: Springer.

Dellaert, B., A. Borgers, and H. Timmermans. 1995. A day in the city: Using conjoint choice experiments to model urban tourists' choice of activity packages. *Tourism Management* 16:347–53.

Desarbo, W. S., V. Ramaswamy, and S. H. Cohen. 1995. Market segmentation with choice-based conjoint analysis. *Marketing Letters* 6(2):137–47.

Dey, M., and R. Mukerjee. 1999. *Fractional Factorial Plans*. New York: Wiley.

Finkbeiner, C. T. 1988. Comparison of conjoint choice simulators. Paper presented at the Sawtooth Software Conference on Perceptual Mapping, Conjoint Analysis, and Computer Interviewing, 2:75–103, Ketchum, ID.

Finney, D. J. 1971. *Probit Analysis*. Cambridge: Cambridge University Press.

Fowkes, T., and M. Wardman. 1988. The design of stated preference travel choice experiments. *Journal of Transport Economics and Policy* 13(1):27–44.

Green, P. E., and A. M. Krieger. 1988. Choice rules and sensitivity analysis in conjoint simulators. *Journal of the Academy of Marketing Science* 6:114–27.

Green, P. E., and A. M. Krieger. 1990. A hybrid conjoint model for price-demand estimation. *European Journal of Operational Research* 44:28–38.

Green, P. E., and A. M. Krieger. 1991. Segmenting markets with conjoint analysis. *Journal of Marketing* 55:20–31.

Green, P. E., A. M. Krieger, and Y. Wind. 2001. Thirty years of conjoint analysis: Reflections and prospects. *Interfaces* 31(3):S56–S73.

Green, P. E., and V. Srinivasan. 1978. Conjoint measurement in consumer research: Issues and outlook. *Journal of Consumer Research* 5:103–120.

Green, P. E., and V. Srinivasan. 1990. Conjoint analysis in marketing research: New developments and directions. *Journal of Marketing* 54:3–19.

Grossmann, H., U. Grasshoff, H. Holling, and R. Schwabe. 2004. Optimal designs for main effects in linear paired comparison models. *Journal of Statistical Planning and Inference* 126:361–76.

Grossmann, H., H. Holling, U. Grasshoff, and R. Schwabe. 2003. A comparison of efficient designs for choices between two options. *Indian Journal of Agricultural Statistics* 56:302–314.

Grossmann, H., H. Holling, and R. Schwabe. 2002. Econometric models in marketing: Advances in optimum experimental design for conjoint analysis and discrete choice models. In *Advances in econometrics*, ed. P. H. Franses and A. L. Montgomery, 93–117. Amsterdam: JAI Press.

Gustafsson, A., A. Herrmann, and F. Huber. 2007. *Conjoint measurement, methods and applications,* 4th edition. New York: Springer.

Haider, W., and G. O. Ewing. 1990. A model of tourist choices of hypothetical Caribbean destination. *Leisure Sciences* 12:33–47.

Hedayat, A. S., N. J. A. Sloane, and J. Stufken. 1999. *Orthogonal arrays: Theory and applications*. New York: Springer.

Hensher, D. A. 1989. Urban tollways and the valuation of travel time savings. *Economic Record* 66:146–56.

Hoffman, P. J. 1960. The paramorphic representation of clinical judgment. *Psychological Bulletin* 47:116–31.

Hoffman, P. J. 1968. Cue-consistency and configurality in human judgment. In *Formal representation of human judgment*, ed. B. Kleinmuntz, 53–90. New York: Wiley.

Hoffman, P. J., P. Slovic, and L. G. Rorer. 1968. An analysis of variance model for the assessment of configural cue utilization in clinical judgment. *Psychological Bulletin* 69:338–49.

Horton, R. W., and B. S. Meldrum. 1973. Seizures induced by allylglycine, 3-mercaptopropionic acid and 4-deoxypyridoxine in mice and photosensitive baboons and different modes of inhibition of cerebral glutamic acid deearboxylase. *British Journal of Pharmacology* 49:52–63.

Huber, J., and C. Puto. 1983. Market boundaries and product choice: Illustrating attraction and substitution effects. *Journal of Consumer Research* 10:34–44.

Johnson, R. M., and K. A. Olberts. 1991. Using conjoint analysis in pricing studies: Is one price variable enough? *American Marketing Association Advanced Research Technique Forum Conference Proceedings* volume (4), 164–73.

Kamakura, W. A., and R. K. Srivastava. 1984. Predicting choice shares under conditions of brand interdependence. *Journal of Marketing Research* 21:420–34.

Katoshevski, R., and H. Timmermans. 2001. Using conjoint analysis to formulate user-centred guidelines for urban design: The example of new residential development in Israel. *Journal of Urban Design* 6:37–53.

Khuri, A. I., and J. A. Cornell. 1987. *Response surface: Design and analysis*. New York: Dekker.

Koelemeijer, K., and H. Oppewal. 1999. Assessing the effects of assortment and ambience: A choice experimental approach. *Journal of Retailing* 75(3, Fall):319–45.

Krantz, D. H. 1964. Conjoint measurement: The Luce-Tukey axiomatization and some extensions. *Journal of Mathematical Psychology* 1:248–77.

Kreher, D. 1996. t-Designs, $t \geq 3$. In *Handbook of combinatorial designs,* ed. C. J. Colbourn and J. H. Dinitz, 47–66. Boca Raton, FL: CRC Press.

Lancaster, K. J. 1966. A new approach to consumer theory. *The Journal of Political Economy* 74:132–57.

Landis, J. R., and G. G. Koch. 1977. The measurement of observer agreement for categorical data. *Biometrics* 33:159–74.

Lazari, A. G., and D. A. Anderson. 1994. Designs of discrete choice set experiments of estimating both attribute and availability cross effects. *Journal of Marketing Research* 31:375–83.

Louviere, J. 1988. *Analyzing decision making: Metric conjoint analysis*. Newbury Park, CA: Sage.

Louviere, J., D. Hencher, and J. Swait. 2000. *Stated choice methods: Analysis and applications*. New York: Cambridge University Press.

Luce, R. D., 1959. *Individual choice behavior: A theoretical analysis.* New York: Wiley.

Luce, R. D., and J. W. Tukey 1964. Simultaneous conjoint measurement: A new type of fundamental measurement. *Journal of Mathematical Psychology* 1:1–27.

McFadden, D. 1974. Conditional logit analysis of qualitative choice behavior. In *Frontiers in econometrics*, ed. P. Zarembka, 105–142. New York: Academic Press.

Montgomery, D. C. 1991. Design and analysis of experiments. New York: Wiley.

Moore, W. L., J. J. Louviere, and R. Verma. 1999. Using conjoint analysis to help design product platforms. *Journal of Product Innovation Management* 16(1, January):27–39.

Morrison, D. F. 1967. *Multivariate statistical methods.* New York: McGraw-Hill.

Oppewal, H., J. J. Louviere, and H. J. P. Timmermans. 1994. Modeling hierarchical conjoint processes with integrated choice experiments. *Journal of Marketing Research* 31:92–105.

Oppewal, H., H. J. P. Timmermans, and J. J. Louviere. 1997. Modelling the effects of shopping centre size and store variety on consumer choice behaviour. *Environment and Planning A* 29(6):1073–90.

Orme, B. K., and W. C. King. 2006. *Getting started with conjoint analysis: Strategies for product design and pricing research.* Madison, WI: Research Publishers.

Piepel, G. F., and J. A. Cornell. 1985. Models for mixture experiments when the response depends on the total amount. *Technometrics* 27:219–27.

Propper, C. 1995. The disutility of time spent on the United Kingdom's National Health Service waiting lists. *Journal of Human Resources* 30:677–700.

Pullman, M. E., K. J. Dodson, and W. L. Moore. 1999. A comparison of conjoint methods when there are many attributes. *Marketing Letters* 10:125–38.

Raghavarao, D. 1971. *Constructions and combinatorial problems in design of experiments.* New York: Wiley.

Raghavarao, D. 1983. *Statistical techniques in agricultural and biological research.* New Delhi: Oxford IBM.

Raghavarao, D., W. T. Fédérer, and S. J. Schwager. 1986. Characteristics for distinguishing among balanced incomplete block designs with repeated blocks. *Journal of Statistical Planning and Inference* 13:151–63.

Raghavarao, D., and L. V. Padgett. 2005. *Block designs: Analysis, combinatorics and applications.* Hackensack, NJ: World Scientific.

Raghavarao, D., and J. B. Wiley. 1987. Testing competing effects among soft drink brands. In *Statistical design theory and practice*, ed. C. E. McCulloch, S. J. Schwayer, G. Casella, and S. R. Searle, 161–76. Ithaca, NY: Cornell University Press.

Raghavarao, D., and J. B. Wiley. 1994. Experimental designs for availability effects and cross effects with one attribute. *Communications in Statistics* 23:1835–46.

Raghavarao, D., and J. B. Wiley. 1998. Estimating main effects with Pareto optimal subsets. *Australian and New Zealand Journal of Statistics* 40:425–32.

Raghavarao, D., and J. B. Wiley. 2006. Design strategies for sequential choice experiments involving economic alternatives. *Journal of Statistical Planning and Inference* 136:3287–3306.

Raghavarao, D., and J. B. Wiley. 2008. Some results on availability designs using Hadamard matrices. *Utilitas Mathematica* 77:35–40.

Raghavarao, D., and J. B. Wiley. 2009a. Conjoint measurement with constraints on attribute levels: A mixture-amount model approach. *International Statistical Review* 77:167–78.

Raghavarao, D., and J. B. Wiley. 2009b. Availability effects (specific combining effects) and interactions. *Sankhya* 71B:175–86.

Raghavarao, D., and D. Zhang. 2002. 2^n behavioural experiments using Pareto optimal choice sets. *Statistica Sinica* 12 1085–92.

Raghavarao, D., and B. Zhou. 1998. Universal optimality of UE 3-designs for a competing effects model. *Communications in Statistics* 27:153–64.

Raktoe, B. L., A. Hedayat, and W. T. Federer. 1981. *Factorial designs*. New York: Wiley.

Ryan, M., and S. Farrar. 2000. Using conjoint analysis to elicit preferences for health care. *British Medical Journal* 320:1530–33.

Ryan, M., E. McIntosh, and P. Shackley. 1998. Methodological issues in the application of conjoint analysis in health care. *Health Economics* 7:373–78.

SAS. 2008. Cary, NY: SAS Institute.

Sawtooth Software. 2009. Sequim, WA: Sawtooth Software.

Scheffé, H. 1958. Experiments with mixtures. *Journal of the Royal Statistical Society Series B* 20:344–60.

Scheffé, H. 1963. The simplex-centroid design for experiments with mixtures. *Journal of the Royal Statistical Society Series B* 25:235–51.

Schwabe, R., U. Grasshoff, H. Grossman, and H. Holling. 2002, Optimal paired comparison designs for partial profiles. *Proceedings of the Fourth International Conference on Mathematical Statistics* 26:79–86.

Simon, H. A. 1957. *Models of man: Social and rational. Mathematical essays on rational human behavior in a social setting*. New York: Wiley.

Sniderman, P. M. and D. B. Grob. 1996. Innovations of experimental design and attitude surveys. *Annual Review of Sociology* 22: 377–399.

Sprott, D. A. 1954. A note on balanced incomplete block designs. *Canadian Journal of Mathematics* 6:341–46.

SPSS. 2008. Chicago: SPSS.

Stat-Ease. 2008. Minneapolis, MN: Stat-Ease.

Street, A. P., and D. J. Street. 1987. *Combinatorics of experimental design*. Oxford, NY: Oxford Science.

Street, D., and L. Burgess. 2007. *The construction of optimal stated choice experiments: Theory and methods*. Hoboken, NJ: Wiley.

Thurstone, L. L. 1927. A law of comparative judgment. *Psychological Review* 34:273–86.

Torgerson, W. S. 1958. *Theory and methods of scaling*. Oxford, UK: Wiley.

Train, K. 2003. *Discrete choice methods with simulation*. New York: Cambridge University Press.

Tversky, A. 1967. A general theory of polynomial CA measurement. *Journal of Mathematical Psychology* 4:1–20.

Van der Waerden, P., H. Oppewal, and H. Timmermans 1993. Adaptive choice behaviour of motorists in congested shopping centre parking lots. *Transportation* 20:395–408.

Wiley, J. B. 1978. Selecting Pareto optimal subsets from multi-attribute alternatives. In *Advances in consumer research*, ed. K. Hunt, 5:171–74.

Wiley, J. B. 1997. Head-to-head: A restaurant decides whether to increase prices. *Journal of Food Products Marketing* 4(2):49–64.

Wiley, J. B., and J. T. Low. 1983. A Monte Carlo simulation study of two approaches for aggregating conjoint data. *Journal of Marketing Research* 20:405–416.

Wiley, J. B., D. Raghavarao, and J. Hargreaves. 1995. Dominance effects in choice based conjoint analysis. Paper presented at the Marketing Science Meetings, Sydney.

Wittink, D. R., and P. Cattin. 1982. Commercial use of conjoint analysis: A Survey. *Journal of Marketing* 46:44–53.

Wittink, D. R., and P. Cattin. 1989. Commercial use of conjoint analysis: An update. *Journal of Marketing* 53:91–96.

Zhang, D. 2001. Pareto optimal design in behavioral experiments. PhD dissertation, Temple University, Philadelphia.

Index

D

E

F

H

I

L

M